T0145090

Studies in Computational Intelligence

Volume 487

Series Editor

J. Kacprzyk, Warsaw, Poland

For further volumes:
http://www.springer.com/series/7092

Cai-Nicolas Ziegler

Social Web Artifacts for Boosting Recommenders

Theory and Implementation

 Springer

PD Dr. Cai-Nicolas Ziegler
PAYBACK GmbH (American Express)
Albert-Ludwigs-Universität Freiburg i.Br.
München
Germany

ISSN 1860-949X   ISSN 1860-9503   (electronic)
ISBN 978-3-319-03287-0   ISBN 978-3-319-00527-0 (eBook)
DOI 10.1007/978-3-319-00527-0
Springer Cham Heidelberg New York Dordrecht London

Printed on acid-free paper

Springer is part of Springer Science+Business Media (www.springer.com)

Foreword

I first met Dr. Ziegler when he was a Ph.D. student spending a few months visiting our GroupLens Research lab in Minnesota. From the first I could tell he was a researcher of unusual vision, not content to work within the bounds of the previous literature on recommenders, but wanting to understand how the early recommender tools could be reshaped to meet needs that their users didn't even imagine they had yet. He was particularly interested in understanding the fit between recommenders as these magical devices that were expected to surprise and delight their users, and the users' real information needs across a variety of interests.

Dr. Ziegler's time with GroupLens was fruitful, producing an excellent paper that characterized the problem of *topic diversification*. The key insight in this work is that even once a recommender algorithm has zoomed in accurately on a user's interests, providing a set of results composed of the items that are individually predicted to be the most interesting may lead to a bored user. For instance, a user who loves *Star Trek* movies might like to have one new *Star Trek* movie recommended, but will certainly be disappointed to have a list containing only the ten most recent *Star Trek* movies. Paradoxically, the recommender must recommend items that are less preferred in order to produce a list that is more preferred. Dr. Ziegler's paper defined this problem carefully, gave a firm mathematical foundation rich enough to support a variety of approaches to diversification, and demonstrated that in practice users prefer the more diverse lists.

This work is therefore characteristic of Dr. Ziegler. He found a core problem that was poorly understood, gave it a strong foundation, and helped the community see its importance. His paper on this mentioned topic is highly cited in the research literature to this day.

This book contains a rich set of examples of this research approach in practice, in several key domains. In addition to a framing of the recommendation problem, there are three deep contributions of the book to a richer understanding of recommendation: *topic diversification* (which we have already discussed), *taxonomy-driven filtering*, and *trust models*.

The key idea behind *taxonomy-driven filtering* is that users often have different levels of interest for different parts of the taxonomy of an information space. For instance, one user who works with the Java programming language may be particularly interested in new work on the type system, while another user may be most interested in Java-based Web containers. Recommenders that are aware of the possibility of these differences can gain predictive power. In the early days of recommender systems taxonomies such as these were not available to use, even if the algorithmic tools had been available. A key contribution of the book is to demonstrate that today there are two rich sources of such information. First, several Social Web projects are creating large, open information taxonomies, such as the category hierarchy in Wikipedia. Second, powerful methods of text processing enable the automatic extraction of taxonomies from textual information spaces. Looking to the future, we can predict that such tools will soon work for music, photos, and even movies. Mining and making use of these taxonomies opens the potential for powerful new approaches to recommendation.

The key idea behind *trust models* is that humans have notions of trust that are not always compatible with the recommendations from a "black box" recommender. Exposing how the recommender model works, and, crucially, exposing which other *humans* have contributed to a set of recommendations can have a big influence on how much the recipient of the recommendations trust them. This book explores trust models based on homophily between members of a recommender community. Over the long term we can expect trust models like these that cross communities, that can be manipulated by the end user ("no, I don't trust that guy!"), and that provide explanations for why a recommendation can be trusted ("your friends Alice and Beth both liked this quadricopter, so you probably will too").

Dr. Ziegler is a visionary scientist, and this book demonstrates his keen insight to new approaches to thinking about recommendation that are now being explored by hundreds of other scientists worldwide. In reading this book you will engage with important problems in recommendation, and will see how thinking deeply about user needs leads to fresh insights into technological possibilities.

Minneapolis, USA   John Riedl
March 2013

Preface

Recommender systems, those software programs that learn from human behavior and make predictions of what products or services we are expected to appreciate and thus purchase, have become an integral part of our everyday life. They proliferate across electronic commerce around the globe. Take, for instance, Amazon.com, the first commercially successful and prominent example of such systems, making use of a broad range of recommender system types: The company is said to have experienced significant double-digit growth in sales solely through personalization, thus representing an impressive uplift. Numerous systems have followed and today, recommender systems exist for virtually all sorts of consumable goods, e.g., books, movies, music, and even jokes.

At the same time, a new evolution on the Web has started to take shape, known as "participation age", "collective wisdom", and – most widely used today – "Web 2.0" or "Social Web": Consumer-generated media and content has become rife, social networks have emerged and are pulling significant shares of the overall Web traffic. In line with these developments, novel information and knowledge structures have become readily available on the Web: People's personal ties and trust links, human-crafted large taxonomies for organizing and categorizing all kinds of items. For example, the massive DMOZ Open Directory Project that has taken on the challenge to categorize the entire Web by its classification system.

This textbook presents approaches to exploit the new Social Web fountain of knowledge, zeroing in first and foremost on two information artifacts, namely classification taxonomies and trust networks. These two are used to improve the performance of product-focused recommender systems: While classification taxonomies are appropriate means to fight the sparsity problem prevalent in many productive recommender systems, interpersonal trust ties – when used as proxies for interest similarity – are able to mitigate the recommenders' scalability problem.

While maintaining the principal focus of improving product recommender systems through taxonomies and trust, several digressions from this main theme are included, such as the use of Web 2.0 taxonomies for computing the semantic proximity of named entity pairs, or the recommending of technology synergies based on Wikipedia and our semantic proximity framework. These slight digressions, however, make the book even more valuable by adding perspectives of what else can be achieved with those precious instruments of knowledge that can be created from the Web 2.0's overtly accessible raw material of data and information.

München, Germany   Cai-Nicolas Ziegler
March 2013

Acknowledgements

Most of the research presented in this textbook has been conducted during my Ph.D. period at the Albert-Ludwigs-Universität Freiburg i.Br., Germany, as well as Group-Lens Research at the University of Minnesota, USA.

Above all, I would like to thank Prof. Dr. Georg Lausen, my supervisor at DBIS, the Institute of Databases and Information Systems in Freiburg. He has been my mentor throughout my Ph.D. period, and has continued to be so ever since. I owe him a lot and value him not only for his work, but also for the person he is.

I would also like to thank my second supervisor, Prof. Dr. Joseph A. Konstan, as well as Prof. Dr. John Riedl, both from the GroupLens Research lab in Minneapolis. These two subject matter experts have provided fresh new input from a different, more HCI-focused perspective, which makes this book even more valuable to the reader.

A big thanks also goes to Prof. Dr. Dr. Lars Schmidt-Thieme, who has introduced me to methods of quantifying the performance of recommender systems in offline experiments. It is now many years ago that I first came to his office in order to discuss collaborative filtering. And that I came out of it with a wealth of new insights and knowledge.

My gratitude is expressed also to the researchers who helped me along the way, particularly Dr. Paolo Massa, Zvi Topol, Ernesto Díaz-Avilés, Prof. Dr. Jennifer Golbeck, Dr. Sean M. McNee, Prof. Dr. Dan Cosley, Dr. Maximilian Viermetz, and Dr. Stefan Jung. It goes on to Ron Hornbaker and Erik Benson, maintaining the All Consuming and BookCrossing community, respectively. They provided the community data for rendering the online user studies possible.

Upon covering the research side of contributions, I now switch to the more emotional ones: Namely my family, my parents Klaus and Angelika, who have always been there for me. As well as my "little" brother Chris. And for sure my beloved wife Miriam, the best that ever happened to me in my life.

To my parents, and Chris, my "little" brother.

Contents

Abbreviations

ANOVA	Analysis of Variance
BLRT	Binomial Log-Likelihood Ratio Test
CBF	Content-Based Filtering
CF	Collaborative Filtering
CGM	Consumer-Generated Media
FOAF	Friend of a Friend
HCI	Human-Computer Interaction
HITS	Hypertext-Induced Topic Search
IDF	Inverse Document Frequency
IR	Information Retrieval
MAE	Mean Absolute Error
ML	Machine Learning
MSE	Mean Squared Error
NDPM	Normalized Distance-Based Performance Measure
ODP	Open Directory Project
PKI	Public Key Infrastructure
RDBMS	Relational Database Management System
RDF	Resource Description Framework
ROC	Receiver Operating Characteristic
SNA	Social Network Analysis
SVD	Singular Value Decomposition
SVM	Support Vector Machine
TF	Term Frequency
VSM	Vector Space Model

Part I
Laying Foundations

Chapter 1
Introduction

"If the doors of perception were cleansed, every thing would appear to man as it is, infinite."

– William Blake (1757 – 1827)

1.1 Motivation

We are living in an era abounding in data and information. And we human beings are ever-hungry for acquiring knowledge. However, the sheer masses of information surrounding us are making the task of dissecting noise from signal, or relevant from irrelevant information, virtually impossible.

Consequently, since more than 60 years [van Rijsbergen, 1975], significant research efforts have been striving to conceive automated filtering systems that provide humans with desirable and relevant information only. Search engines count among these filtering systems and have gained wide-spread acceptance, rendering information search feasible even within chaotic and anarchical environments such as the Web.

During the last 15 years, recommender systems [Resnick and Varian, 1997; Konstan, 2004] have been gaining momentum as another efficient means of reducing complexity when searching for relevant information. Recommenders intend to provide people with suggestions of products they will appreciate, based upon their past preferences, history of purchase, or demographic information [Resnick et al, 1994].

1.1.1 Collaborative Filtering Systems

Most successful industrial and academic recommender systems employ so-called collaborative filtering techniques [Goldberg et al, 1992]. Collaborative filtering systems mimic social processes in an automated fashion: In the "real world" we ask like-minded friends or family members for their particular opinion on new book

C.-N. Ziegler: *Social Web Artifacts for Boosting Recommenders*, SCI 487, pp. 3–9.
DOI: 10.1007/978-3-319-00527-0_1 © Springer International Publishing Switzerland 2013

releases. In the virtual space, recommender systems do this job for us. Their principal mode of operation can be broken down into three major steps:

- **Profiling.** For each user a_i part of the community, an interest profile for the domain at hand, e.g., books, videos, etc., is computed. In general, these profiles are represented as partial *rating functions* $r_i : B \to [-1, +1]^\perp$, where $r_i(k) \in [-1, +1]$ gives a_i's rating for product b_k, taken from the current domain of interest. Ratings are expressions of value for products. High values $r_i(k) \to +1$ denote appreciation, while low values $r_i(k) \to -1$ indicate dislike.
- **Neighborhood formation.** Neighborhood formation aims at finding the best-M like-minded neighbors of a_i, based on their profiles of interest. Roughly speaking, the more ratings two users a_i, a_j have in common, and the more their corresponding ratings $r_i(k), r_j(k)$ have similar or identical values, the higher the similarity between a_i and a_j.
- **Rating prediction.** Predictions for products b_k still unknown to a_i depend on mainly two factors. First, the *similarity* of neighbors a_j having rated b_k, and second, the *rating* $r_j(k)$ they have assigned to product b_k. Eventually, top-N recommendation lists for users a_i are compiled based upon these predictions.

Hence, the intuition behind collaborative filtering is that if user a_i has agreed with his neighbors in the past, he will do so in the future as well.

1.1.2 The Participation Age on the Web

At the same time as recommender systems have been gaining momentum, the Web 2.0 has changed the Web and its content production process forever:

Instead of a small number of producers that generate large quantities of content, such as news portals, company websites, now *every person* has become a producer and consumer of information on the Web *at the same time*. This hermaphrodite is dubbed "prosumer" [Ritzer and Jurgenson, 2010]. Weblogs, newsgroups, consumer-generated media (CGM) like product ratings in e-commerce portals, Wikipedia, etc. represent the brave new Web: Everybody expresses his own opinion on everything, social ties are made explicit in a digital fashion through popular social network platforms, and millions of humans contribute to create artefacts of collective wisdom, like DMOZ, also known as the Open Directory Project (*http://www.dmoz.org*).

The traces of these Web 2.0 creations can be put to use for making recommender systems perform better and run faster. In this textbook, we investigate the use of mainly two types of information structures that can be harvested from this "Social Web"[1]:

Classification taxonomies. Most renowned for Web 2.0 classification systems are the so-called "folksonomies" (see, e.g., [Halpin et al, 2007] and [Jäschke et al, 2007]). They are built by the collective efforts of users of the Web and not controlled by some central authority. They became popular around 2004 in particular

---

[1] Throughout the book, we use the terms "Web 2.0" and "Social Web" interchangeably.

in software applications such as social bookmarking (see, e.g., *http://del.icio.us*) and annotation of photographs, such as Flickr (see *http://www.flickr.com*). However, folksonomies *are* different from taxonomies: Folksonomies are built on a discrete order, meaning there are just the categories (called "tags" in the context of folksonomies) but no relationship between those. Not so in a taxonomy, where its elements are linked by hierarchical relationships, so that any category may have up to one super-concept and zero or more sub-concepts. At the same time, efforts have started gaining momentum where even *taxonomies* are crafted by collaborative efforts of thousands of people on the Social Web. The huge advantage of their collective nature is that these taxonomies may become much larger than their centrally controlled peers, tying in much more human resource capital to their genesis. These Web 2.0 taxonomies, best represented by the before mentioned DMOZ, have been found to be excellent means for use in information filtering (see, e.g., [Banerjee and Scholz, 2008]). These are also the ends we want to use them for, as background knowledge for improving the *quality* (and thus the value to users) of predictions of recommender systems.

Social ties and trust networks. Another even more remarkable evolution is that of social networks on the Web. And in contrast to taxonomies, which have existed *before* the advent of the Web 2.0, these social networks have emerged *along* with the Web 2.0. There are numerous popular social networks out there. Facebook (*http://www.facebook.com*) now is the most prominent one, boasting more than one billion users[2]. Other large networks are LinkedIn (*http://www.linkedin.com*) or Myspace (*http://www.myspace.com*), which used to be the biggest social network but lost tremendous market share to Facebook. Next to these large social networks, there are numerous substantially smaller ones, such as the communities used in the research conducted in this book, namely All Consuming (*http://www.allconsuming.net*)[3], Advogato (*http://www.advogato.org*), Epinions (*http://www.epinions.com*), and BookCrossing (*http://www.bookcrossing.com*). All these social networks have in common that their users explicitly specify social ties, be it the friends they have or the people they trust. In this book we focus on *trust* networks formed in the online space. By providing empirical evidence that interpersonal trust and attitudinal similarity (i.e., personal interests) correlate, we come to conclude that trust is an effective means to help in the recommendation process: Instead of having complex computations that aim at *finding* the users that are similar to our own behavior, we resort to the explicit trust links that the user has created himself. In this vein, trust ties allow us to reduce the *computation complexity* of recommendation generation and make it more *scalable*.

These two types of information structures, taxonomies and trust networks, are at the heart of all the book's content.

---

[2] This number of users has been reached in October 2012.

[3] Note that the All Consuming community platform that has been the basis of many of our studies presented in this book has changed dramatically in its shape and functionality.

1.2 Organization and Content

Our textbook is organized into four parts. The structure as well as the interdependencies between the various parts and chapters are visualized in Figure 1.1.

The first part comprises of the chapter at hand, as well as a general introduction into recommender systems.

The second part is all about the use of taxonomies in information filtering: Chapter 3 is hereby the central core, introducing our taxonomy-driven method for computing product recommendations. This method also includes a novel approach for *diversifying* recommendation lists, which is assessed with rigor and more detail in Chapter 4. Chapter 5 again builds upon the taxonomy-driven filtering approach and introduces a framework that allows to calculate the semantic proximity of named entity pairs. This semantic proximity framework lies at the heart of our technology synergy recommender, presented in Chapter 6.

Part III abandons taxonomies and recommendations for a while and analyzes *trust* as a form of social tie. The focus of Chapter 7 is on trust *networks* and how trust can be propagated along edges within these. Chapter 8 identifies correlations between trust and interest similarity, both by working on literature from social psychology as well as by presenting empirical evaluations on the Web, and thus paves the way for using trust as suitable proxy in finding similar peers within generic collaborative filtering frameworks.

Having introduced trust and taxonomies as information structures for improving the recommendation generation process, we are putting these two ingredients together in Part IV: Chapter 9 presents our approach to *decentralized* recommender systems, making use of the contributions outlined before in Part II and III. Chapter 10 concludes this book and ventures into the nearer future.

1.3 Contributions

We concentrate on how data and information structures prevalent on the Social Web can help to improve recommender systems; either by improving the accuracy and quality of the recommendations made, or by making them more scalable in terms of computation complexity.

The contributions made in this book refer to various disciplines, e.g., information filtering and retrieval, (social) network analysis, human-computer interaction, and social psychology. All these building bricks are contributions in their own right and revolve around the two dominating themes laid out before.

- **Taxonomy-driven filtering.** Taxonomy-driven filtering relies upon very large product classification taxonomies as powerful background knowledge to render recommendation computations feasible in environments where sparsity prevails. Our proposed approach features two important contributions:

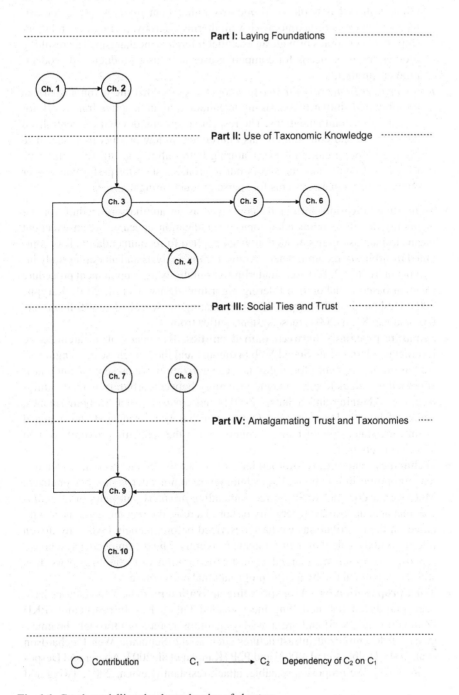

Fig. 1.1 Graph modelling the dependencies of chapters

Interest profile assembly and similarity measurement. Our method for assem
bling profiles based on *topic interests* rather than *product ratings* constitutes the nucleus of taxonomy-driven filtering. Hereby, hierarchical relationships between topics provide an essential inference mechanism. The resulting
profiles serve as means for computing user-user, user-product, and product-
product similarity.

Recommender framework for sparse data. We embedded the taxonomy-driven
profiling and similarity measuring technique into an adaptive framework for
computing recommendations. The respective system, devised for centralized
data storage and computation, also introduces a new product-user relevance
predictor. Empirical evidence, featuring both online and offline evaluations
for two different datasets, shows our approach's superior performance over
common benchmark systems for sparse product-rating matrices.

- **Topic diversification.** Originally conceived as an auxiliary procedure for the
taxonomy-driven recommender, topic diversification appeared as an excellent
means to increase user satisfaction when applied to recommendation lists computed by arbitrary recommenders. Ample offline analysis and an online study involving more than 2,100 users underline its suitability as a supplement procedure
for item-based collaborative filtering algorithms [Sarwar et al, 2001; Karypis,
2001; Deshpande and Karypis, 2004], mitigating the so-called "portfolio effect"
[Ali and van Stam, 2004] these systems suffer from.

- **Semantic proximity between named entities.** By means of an architecture
leveraging a large-scale Social Web taxonomy and the Google search engine, as
well as our novel metric for multi-class categorization, we present an automated
approach to calculate the semantic proximity between arbitrary named entities
(see, e.g., Manning and Schütze [2001]) and concepts, such as brand names,
names of people, locations, etc. Our system has been evaluated and matched
against the statements of human arbiters, rating the semantic proximity of two
sets of concept pairs.

- **Technology synergies recommender.** In contrast to all other recommender systems proposed in this book, the technology recommender is *non*-personalized.
Moreover, it does not focus on recommending *products*, but pairs of *technologies* that are supposedly synergetic (or not). Hereby, the recommender is heavily
based on the contributions we have described before, namely taxonomy-driven
filtering and the calculation of semantic proximity. The recommender system we
propose has been benchmarked against expert ratings of technology pairs. It is
still in operational use by a large multi-national corporation.

- **Trust propagation based on spreading activation models.** Trust metrics have
been introduced for modelling the so-called Public Key Infrastructure (PKI)
[Zimmermann, 1995] and are nowadays gaining momentum through the emergence of decentralized infrastructures such as the Semantic Web [Richardson
et al, 2003; Golbeck et al, 2003] and P2P [Kamvar et al, 2003; Aberer and Despotovic, 2001]. We propose a scalable, attack-resistant [Levien, 2004; Twigg and

Dimmock, 2003] trust metric based upon spreading activation models which is able to compute *neighborhoods* of most-trustworthy peers.

- **Interpersonal trust and interest similarity.** While the proverbial saying that "birds of a feather flock together" suggests that bonds of trust are mostly forged among like-minded individuals, no empirical evidence has been given so far. Social psychology has conducted considerable research on interactions between interpersonal *attraction* and attitudinal similarity [Berscheid, 1998], but not with respect to interpersonal *trust*. We therefore present two frameworks for analyzing interdependencies between trust and similarity and provide empirical, statistically significant evidence from two communities: First, from an online book-reading community, second, from a community catering to film lovers. Evaluations in both communities show that trust and similarity *do* correlate within this context.

- **Decentralized recommender framework.** Our final contribution aims at the seamless integration of many of the previous results and techniques into one coherent framework for decentralized recommendation making. Empirical results based on offline analysis are given in order to compare its efficiency with two non-decentralized recommenders not making use of trust-based neighborhood formation.

1.3.1 Publications

The contributions described in this book have been published in international conferences [Ziegler and Lausen, 2004c,a,b; Ziegler et al, 2004a, 2005, 2006; Ziegler and Jung, 2009; Ziegler and Viermetz, 2009], refereed workshops [Ziegler, 2004a; Ziegler et al, 2004b; Ziegler, 2004b], book chapters [Ziegler, 2009] and journals [Ziegler and Lausen, 2005; Ziegler and Golbeck, 2007; Ziegler and Lausen, 2009].

Chapter 2
On Recommender Systems

"The truth is incontrovertible. Malice may attack it, ignorance may deride it, but in the end, there it is."

– Winston Churchill (1874 – 1965)

2.1 Introduction

Recommender systems [Resnick and Varian, 1997] have gained wide-spread acceptance and attracted increased public interest during the last decade, levelling the ground for new sales opportunities in e-commerce [Schafer et al, 1999; Sarwar et al, 2000a]. For instance, online retailers like Amazon.com (*http://www.amazon.com*) successfully employ an extensive range of different types of recommender systems.

Their principal objective is that of complexity reduction for the human being, sifting through very large sets of information and selecting those pieces that are relevant for the active user[1]. Moreover, recommender systems apply personalization techniques, considering that different users have different preferences and different information needs [Konstan et al, 1997]. For instance, supposing the domain of book recommendations, historians are supposedly more interested in medieval prose, e.g., Geoffrey Chaucer's Canterbury Tales, than literature about self-organization, which might be more relevant for AI researchers.

2.2 Collecting Preference Information

Hence, in order to generate personalized recommendations that are tailored to the active user's specific needs, recommender systems must collect personal preference information, e.g., the user's history of purchase, click-stream data, demographic information, and so forth. Traditionally, expressions of preference of users a_i for

---

[1] The term "active user" refers to the person for whom recommendations are made.

C.-N. Ziegler: *Social Web Artifacts for Boosting Recommenders*, SCI 487, pp. 11–20.
DOI: 10.1007/978-3-319-00527-0_2   © Springer International Publishing Switzerland 2013

products b_k are generally called *ratings* $r_i(b_k)$. Two different types of ratings are distinguished:

Explicit ratings. Users are required to *explicitly* specify their preference for any particular item, usually by indicating their extent of appreciation on 5-point or 7-point likert scales. These scales are then mapped to numeric values, for instance continuous ranges $r_i(b_k) \in [-1, +1]$. Negative values commonly indicate dislike, while positive values express the user's liking.

Implicit ratings. Explicit ratings impose additional efforts on users. Consequently, users often tend to avoid the burden of explicitly stating their preferences and either leave the system or rely upon "free-riding" [Avery and Zeckhauser, 1997]. Alternatively, garnering preference information from mere *observations* of user behavior is much less obtrusive [Nichols, 1998]. Typical examples for implicit ratings are purchase data, reading time of Usenet news [Resnick et al, 1994], and browsing behavior [Gaul and Schmidt-Thieme, 2002; Middleton et al, 2004]. While easier to collect, implicit ratings bear some serious implications. For instance, some purchases are gifts and thus do not reflect the active user's interests. Moreover, the inference that purchasing implies liking does not always hold.

Owing to the difficulty of acquiring explicit ratings, some providers of product recommendation services adopt bilateral approaches. For instance, Amazon.com computes recommendations based on explicit ratings *whenever possible*. In case of unavailability, observed implicit ratings are used instead.

2.3   Recommender System Types and Techniques

Two principal paradigms for computing recommendations have emerged, namely *content-based* and *collaborative* filtering [Goldberg et al, 1992]. Content-based filtering, also called *cognitive filtering* [Malone et al, 1987], computes similarities between the active user a_i's basket of appreciated products, and products from the product universe that are still unknown to a_i. Product-product similarities are based on features and selected attributes. Whereas collaborative filtering, also called *social filtering* [Resnick et al, 1994], computes similarities between *users*, based upon their rating profile. Most similar users then serve as "advisers" suggesting the most relevant products to the active user.

Advanced recommender systems tend to *combine* collaborative and content-based filtering, trying to mitigate the drawbacks of either approach and exploiting synergetic effects. These systems have been coined "hybrid systems" [Balabanović and Shoham, 1997]. Burke [2002] provides an extensive survey of hybridization methods.

2.3.1   Content-Based Techniques

Content-based approaches to recommendation making are deeply rooted in information retrieval [Baudisch, 2001]. Typically, these systems learn Bayesian classifiers through content features [Lang, 1995; Ghani and Fano, 2002; Lam et al, 1996; Sollenborn and Funk, 2002], or perform nearest-neighbor vector-space queries [Pazzani, 1999; Alspector et al, 1998; Mukherjee et al, 2001; Ferman et al, 2002]. Bayesian classifiers use Bayes' theorem of conditional independence:

$$P(R \mid F) = \frac{P(F \mid R) \cdot P(R)}{P(F)} \tag{2.1}$$

Moreover, Bayesian classifiers make the "naïve" assumption that product description features are independent, which is usually not the case. Given the class label, the probability of b_k belonging to class R_i, given its n feature values $F_1, \ldots, F_n$, is defined as follows:

$$P(R_i \mid F_1, \ldots, F_n) = \frac{1}{Z} \cdot P(R_i) \cdot \prod_{j=1}^{n} P(F_j \mid R_i) \tag{2.2}$$

Variable Z represents a scaling factor only dependent on $F_1, \ldots, F_n$. Probabilities $P(R_i)$ and $P(F_j \mid R_i)$ can be estimated from training data.

For vector-space queries, attributes, e.g., plain-text terms or machine-readable metadata, are extracted from product descriptions and used for user profiling and product representation. For instance, Fab [Balabanović and Shoham, 1997] represents documents in terms of the 100 words with the highest TF-IDF weights [Baeza-Yates and Ribeiro-Neto, 2011], i.e., the words that occur more frequently in those documents than they do on average.

2.3.2 Collaborative Filtering

Content-based filtering only works when dealing with domains where feature extraction is feasible and attribute information readily available. Collaborative filtering (CF), on the other hand, uses content-less representations and does not face that same limitation. For instance, Jester [Goldberg et al, 2001] uses collaborative filtering to recommend jokes to its users. While content-based filtering considers the descriptive *features* of products, collaborative filtering uses the *ratings* that users assign to products. Hence, CF algorithms typically operate on a set of users $A = \{a_1, a_2, \ldots, a_n\}$, a set of products $B = \{b_1, b_2, \ldots, b_m\}$, and partial rating functions $r_i : B \rightarrow [-1, +1]^\perp$ for each user $a_i \in A$. Negative values $r_i(b_k)$ denote dislike, while positive values express a_i's liking of product b_k. Bottom values $r_i(b_k) = \perp$ indicate that a_i has not rated b_k.

Owing to their high quality output and minimal information requirements, CF systems have become the most prominent representatives of recommender systems. Many commercial vendors, e.g., Amazon.com [Linden et al, 2003] and TiVo [Ali and van Stam, 2004], use variations of CF techniques to suggest products to their customers. Besides simple Bayesian classifiers [Miyahara and Pazzani, 2000; Breese et al, 1998; Lang, 1995; Lam et al, 1996], horting [Aggarwal et al, 1999], and association rule-based techniques [Sarwar et al, 2000a], mainly two approaches have acquired wide-spread acceptance, namely *user-based* and *item-based* collaborative filtering. In fact, the term "collaborative filtering" is commonly used as a synonym for user-based CF, owing to this technique's immense popularity.

The following two sections roughly depict algorithmic implementations of both user-based and item-based CF.

2.3.2.1 User-Based Collaborative Filtering

The Ringo [Shardanand and Maes, 1995] and GroupLens [Konstan et al, 1997] projects have been among the first recommender systems to apply techniques known as "user-based collaborative filtering". Representing each user a_i's rating function r_i as a vector, they first compute similarities $c(a_i, a_j)$ between all pairs $(a_i, a_j) \in A \times A$. To this end, common statistical correlation coefficients are used, typically Pearson correlation [Resnick et al, 1994], and the cosine similarity measure, well-known from information retrieval [Baeza-Yates and Ribeiro-Neto, 1999]. As its name already suggests, the cosine similarity measure quantifies the similarity between two vectors $\mathbf{v_i}, \mathbf{v_j} \in [-1, +1]^{|B|}$ by the cosine of their angles:

$$\text{sim}(\mathbf{v_i}, \mathbf{v_j}) = \frac{\sum_{k=0}^{|B|} v_{i,k} \cdot v_{j,k}}{\left(\sum_{k=0}^{|B|} v_{i,k}^2 \cdot \sum_{k=0}^{|B|} v_{j,k}^2 \right)^{\frac{1}{2}}} \tag{2.3}$$

Pearson correlation, derived from a linear regression model [Herlocker et al, 1999], is similar to cosine similarity, but measures the degree to which a linear relationship exists between two variables. Symbols $\overline{v_i}, \overline{v_j}$ denote the *averages* of vectors $\mathbf{v_i}, \mathbf{v_j}$:

$$\text{sim}(\mathbf{v_i}, \mathbf{v_j}) = \frac{\sum_{k=0}^{|B|} (v_{i,k} - \overline{v_i}) \cdot (v_{j,k} - \overline{v_j})}{\left(\sum_{k=0}^{|B|} (v_{i,k} - \overline{v_i})^2 \cdot \sum_{k=0}^{|B|} (v_{j,k} - \overline{v_j})^2 \right)^{\frac{1}{2}}} \tag{2.4}$$

Either using the cosine similarity measure or Pearson correlation to compute similarities $c(a_i, a_j)$ between all pairs $(a_i, a_j) \in A \times A$, *neighborhoods* $\text{prox}(a_i)$ of top-M most similar neighbors are built for every peer $a_i \in A$. Next, predictions are computed for all products b_k that a_i's neighbors have rated, but which are yet unknown to a_i, i.e., more formally, predictions $w_i(b_k)$ for $b_k \in \{b \in B \mid \exists a_j \in \text{prox}(a_i) : r_j(b) \neq \perp\}$:

$$w_i(b_k) = \overline{r_i} + \frac{\sum_{a_j \in \text{prox}(a_i)} (r_j(b_k) - \overline{r_j}) \cdot c(a_i, a_j)}{\sum_{a_j \in \text{prox}(a_i)} c(a_i, a_j)} \tag{2.5}$$

Predictions are thus based upon weighted averages of deviations from a_i's neighbors' means. For top-N recommendations, a list $P_{w_i} : \{1, 2, \ldots, N\} \rightarrow B$ is computed, based upon predictions w_i. Note that function P_{w_i} is injective and reflects recommendation ranking in *descending* order, giving highest predictions first.

Performance Tuning

In order to make better predictions, various researchers have proposed several modifications to the core user-based CF algorithm. The following list names the most prominent ones, but is certainly not exhaustive:

Inverse user frequency. In information retrieval applications based on the vector-space model, word frequencies are commonly modified by a factor known as the "inverse document frequency" [Baeza-Yates and Ribeiro-Neto, 2011]. The idea is to reduce the impact of frequently occurring words, and increase the weight for uncommon terms when computing similarities between document vectors. Inverse user frequency, first mentioned by Breese et al [1998], adopts that notion and rewards co-votes for less common items much more than co-votes for very popular products.

Significance weighting. The computation of user-user correlations $c(a_i, a_j)$ only considers products that *both* users rated, i.e., $b_k \in (\{b \mid r_i(b) \neq \perp\} \cap \{b \mid r_j(b) \neq \perp\})$. Hence, even if a_i and a_j have co-rated only one single product b_k, they will have maximum correlation if $r_i(b_k) = r_j(b_k)$ holds. Clearly, such correlations, being based upon few data-points only, are not very reliable. Herlocker et al [1999] therefore proposed to *penalize* user correlations based on fewer than 50 ratings in common, applying a significance weight of $s/50$, where s denotes the number of co-rated items. Default voting [Breese et al, 1998] is another approach to address the same issue.

Case amplification. While both preceding modifications refer to the similarity computation process, case amplification [Breese et al, 1998] addresses the *rating prediction* step, formalized in Equation 2.5. Correlation weights $c(a_i, a_j)$ close to one are emphasized, and low correlation weights punished:

$$c'(a_i, a_j) = \begin{cases} c(a_i, a_j)^\rho, & \text{if } c(a_i, a_j) \geq 0 \\ -(-c(a_i, a_j))^\rho, & \text{else} \end{cases} \tag{2.6}$$

Hence, highly similar users have much more impact on predicted ratings than before. Values ρ around 2.5 are typically assumed.

Filtering Agents

Some researchers [Sarwar et al, 1998; Good et al, 1999] have taken the concept of user-based collaborative filtering somewhat further and added *filterbots* as additional users eligible for selection as neighbors for "real" users. Filterbots are automated programs behaving in certain, pre-defined ways. For instance, in the context of the GroupLens Usenet news recommender [Konstan et al, 1997], some filterbots rated Usenet articles based on the proportion of spelling errors, while others focused on text length, and so forth. Sarwar has shown that filterbots can improve recommendation accuracy when operating in sparsely populated CF systems [Sarwar et al, 1998].

2.3.2.2   Item-Based Collaborative Filtering

Item-based CF [Karypis, 2001; Sarwar et al, 2001; Deshpande and Karypis, 2004] has been gaining momentum over the last five years by virtue of favorable computational complexity characteristics and the ability to decouple the model computation process from actual prediction making. Specifically for cases where $|A| \gg |B|$, item-based CF's computational performance has been shown superior to user-based CF [Sarwar et al, 2001]. Its success also extends to many commercial recommender systems, such as Amazon.com's [Linden et al, 2003] and TiVO [Ali and van Stam, 2004].

As with user-based CF, recommendation making is based upon ratings $r_i(b_k)$ that users $a_i \in A$ provide for products $b_k \in B$. However, unlike user-based CF, similarity values c are computed for *items* rather than *users*, hence $c : B \times B \rightarrow [-1, +1]$. Roughly speaking, two items b_k, b_e are similar, i.e., have large $c(b_k, b_e)$, if users who rate one of them tend to rate the other, and if users tend to assign identical or similar ratings to them. Effectively, item-based similarity computation equates to the user-based case when turning the product-user matrix $90°$. Next, neighborhoods $\text{prox}(b_k) \subseteq B$ of top-M most similar items are defined for each b_k. Predictions $w_i(b_k)$ are computed as follows:

$$w_i(b_k) = \frac{\sum_{b_e \in B'_k} \left(c(b_k, b_e) \cdot r_i(b_e) \right)}{\sum_{b_e \in B'_k} |c(b_k, b_e)|}, \tag{2.7}$$

where

$$B'_k := \{ b_e \in B \,|\, b_e \in \text{prox}(b_k) \wedge r_i(b_e) \neq \bot \}$$

Intuitively, the approach tries to mimic real user behavior, having user a_i judge the value of an unknown product b_k by comparing the latter to known, similar items b_e and considering how much a_i appreciated these b_e.

The eventual computation of a top-N recommendation list P_{w_i} follows the user-based CF's process, arranging recommendations according to w_i in descending order.

2.3.3   Hybrid Recommender Systems

Hybrid approaches are geared towards unifying collaborative and content-based filtering under one single framework, leveraging synergetic effects and mitigating inherent deficiencies of either paradigm. Consequently, hybrid recommenders operate on both product rating information *and* descriptive features. In fact, numerous ways for combining collaborative and content-based aspects are conceivable, Burke [2002] lists an entire plethora of hybridization methods. Most widely adopted among these, however, is the so-called "collaboration via content" paradigm [Pazzani, 1999], where content-based profiles are built to detect similarities among users.

Sample Approaches

One of the earliest hybrid recommender systems is Fab [Balabanović and Shoham, 1997], which suggests Web pages to its users. Melville et al [2002] and Hayes and Cunningham [2004] use content information for *boosting* the collaborative filtering process. Torres et al [2004] and McNee et al [2002] propose various hybrid systems for recommending citations of research papers. Huang et al [2002, 2004] use content-based features in order to construct correlation graphs to explore *transitive* associations between users. Model-driven hybrid approaches have been suggested by Basilico and Hofmann [2004], proposing perceptron learning and kernel functions, and by Schein et al [2002], using more traditional Bayesian classifiers.

2.4 Evaluating Recommender Systems

Evaluations of recommender systems are indispensable in order to quantify how *useful* recommendations made by system S_x are compared to S_y over the complete set of users A. *Online* evaluations, i.e., directly asking users for their opinions, are, in most cases, not an option. Reasons are manifold:

Deployment. In order to perform online evaluations, an intact virtual community able to run recommender system services is needed. On the other hand, successfully deploying an online community and making it become self-sustaining is cumbersome and may exceed the temporal scope of most research projects.

Obtrusiveness. Even if an online community is readily available, evaluations cannot simply be performed at will. Many users may regard questionnaires as an additional burden, providing no immediate reward for themselves, and perhaps even decide to leave the system.

Hence, research has primarily relied upon *offline* evaluation methods, which are applicable to datasets containing past product ratings, such as, for instance, the well-known MovieLens and EachMovie datasets, both publicly available.[2] Machine learning cross-validation techniques are applied to these datasets, e.g., hold-out, K-folding, or leave-one-out testing, and evaluation metrics run upon. The following sections give an outline of popular metrics used for offline evaluations. An extensive and more complete survey is provided by Herlocker et al [2004].

2.4.1 Accuracy Metrics

Accuracy metrics have been defined first and foremost for two major tasks: first, to judge the *accuracy* of single predictions, i.e., how much predictions $w_i(b_k)$ for products b_k deviate from a_i's actual ratings $r_i(b_k)$. These metrics are particularly suited for tasks where predictions are displayed along with the product, e.g., annotation in

---

[2] See *http://www.grouplens.org* for EachMovie, MovieLens, and other datasets.

context [Herlocker et al, 2004]. Second, *decision-support* metrics evaluate the effectiveness of helping users to select high-quality items from the set of all products, generally supposing *binary preferences*.

2.4.1.1 Predictive Accuracy Metrics

Predictive accuracy metrics measure how close predicted ratings come to true user ratings. Most prominent and widely used [Shardanand and Maes, 1995; Herlocker et al, 1999; Breese et al, 1998; Good et al, 1999], mean absolute error (MAE) represents an efficient means to measure the statistical accuracy of predictions $w_i(b_k)$ for sets B_i of products:

$$|\overline{E}| = \frac{\sum_{b_k \in B_i} |r_i(b_k) - w_i(b_k)|}{|B_i|} \tag{2.8}$$

Related to MAE, mean squared error (MSE) *squares* the error before summing. Hence, large errors become much more pronounced than small ones. Very easy to implement, predictive accuracy metrics are inapt for evaluating the quality of top-N recommendation lists: users only care about errors for high-rank products. On the other hand, prediction errors for low-rank products are unimportant, knowing that the user has no interest in them anyway. However, MAE and MSE account for both types of errors in exactly the same fashion.

2.4.1.2 Decision-Support Metrics

Precision and recall, both well-known from information retrieval, do not consider predictions and their deviations from actual ratings. They rather judge how *relevant* a set of ranked recommendations is for the active user.

Typically, before using these metrics, K-folding is applied, dividing every user a_i's rated products $b_k \in R_i = \{b \in B \mid r_i(b) \neq \bot\}$ into K disjoint slices of preferably equal size. Folding parameters $K \in \{4, 5, \ldots, 10\}$ are commonly assumed. Next, $K - 1$ randomly chosen slices are used to form a_i's *training set* R_i^x. These ratings then define a_i's profile from which final recommendations are computed. For recommendation generation, a_i's residual slice $(R_i \setminus R_i^x)$ is retained and not used for prediction. This slice, denoted T_i^x, constitutes the *test set*, i.e., those products the recommendation algorithms intend to predict.

Precision, Recall, and F1

Sarwar et al [2000b] present an adapted variant of recall, recording the percentage of test set products $b \in T_i^x$ occurring in recommendation list P_i^x with respect to the overall number of test set products $|T_i^x|$:

$$\text{Recall} = 100 \cdot \frac{|T_i^x \cap \Im P_i^x|}{|T_i^x|} \tag{2.9}$$

Symbol $\Im P_i^x$ denotes the *image* of map P_i^x, i.e., all items part of the recommendation list.

Accordingly, precision represents the percentage of test set products $b \in T_i^x$ occurring in P_i^x with respect to the size of the recommendation list:

$$\text{Precision} = 100 \cdot \frac{|T_i^x \cap \Im P_i^x|}{|\Im P_i^x|} \qquad (2.10)$$

Another popular metric used extensively in information retrieval and recommender systems research [Sarwar et al, 2000b; Huang et al, 2004; Montaner, 2003] is the standard F1 metric. F1 combines precision and recall in one single metric, giving equal weight to both of them:

$$\text{F1} = \frac{2 \cdot \text{Recall} \cdot \text{Precision}}{\text{Recall} + \text{Precision}} \qquad (2.11)$$

Breese Score

Breese et al [1998] introduce an interesting extension to recall, known as weighted recall or Breese score. The underlying idea refers to the intuition that the expected utility of a recommendation list is simply the *probability* of viewing a recommended product that is actually relevant, i.e., taken from the test set, times its utility, which is either 0 or 1 for implicit ratings. Breese furthermore posits that each successive item in a list is less likely to be viewed by the active user with exponential decay. The expected utility of a ranked list P_i^x of products is as follows:

$$H(P_i^x, T_i^x) = \sum_{b \in (T_i^x \cap \Im P_i^x)} \frac{1}{2^{(P_i^{x-1}(b)-1)/(\alpha-1)}} \qquad (2.12)$$

Parameter α denotes the viewing half-life. Half-life is the number of the product on the list such that there is a 50% chance that the active agent, represented by training set R_i^x, will review that product. Finally, the weighted recall of P_i^x with respect to T_i^x is defined as follows:

$$\text{BScore}(P_i^x, T_i^x) = 100 \cdot \frac{H(P_i^x, T_i^x)}{\sum_{k=1}^{|T_i^x|} \frac{1}{2^{(k-1)/(\alpha-1)}}} \qquad (2.13)$$

Interestingly, when assuming $\alpha = \infty$, Breese score is identical to unweighted recall.

Other popular decision-support metrics include ROC [Schein et al, 2002; Melville et al, 2002; Good et al, 1999], the so-called *receiver operating characteristic*. ROC measures the extent to which an information filtering system is able to successfully distinguish between signal and noise. Less frequently used, NDPM [Balabanović and Shoham, 1997] compares two different, weakly ordered rankings.

2.4.2 Beyond Accuracy

Though accuracy metrics are an important facet of usefulness, there are traits of user satisfaction they are unable to capture. Still, non-accuracy metrics have largely been denied major research interest so far and have only been treated as marginally important supplements for accuracy metrics.

2.4.2.1 Coverage

Among all non-accuracy evaluation metrics, coverage has been the most frequently used [Herlocker et al, 1999; Middleton et al, 2004; Good et al, 1999]. Coverage measures the percentage of elements part of the problem domain for which predictions can be made.

For instance, supposing the user-based collaborative filtering approach presented in Section 2.3.2.1, coverage for the entire set of users is computed as follows:

$$\text{Coverage} = 100 \cdot \frac{\sum_{a_i \in A} |\{b \in B \,|\, \exists a_j \in \text{prox}(a_i) : r_j(b) \neq \bot\}|}{|B| \cdot |A|} \qquad (2.14)$$

2.4.2.2 Novelty and Serendipity

Some recommenders produce highly accurate results that are still useless in practice, e.g., suggesting bananas to customers in a grocery store: almost everyone appreciates bananas, so their recommending implies high accuracy. On the other hand, owing to their high popularity, most people *intuitively* purchase bananas upon entering a grocery store. They do not require an additional recommendation since they "already know" [Terveen and Hill, 2001].

Novelty and serendipity metrics thus measure the *non-obviousness* of recommendations made, penalizing "cherry-picking" [Herlocker et al, 2004].

Part II
Use of Taxonomic Knowledge

Chapter 3
Taxonomy-Driven Filtering

"Recommend to your children virtue; that alone can make them happy, not gold."

– Ludwig van Beethoven (1770 – 1827)

3.1 Introduction

One of the primary issues that recommender systems are facing is rating sparsity, resulting in a decrease of the recommendations' accuracy. Hence, high-quality product suggestions are only feasible when information density is high, i.e., large numbers of users voting for small numbers of items and issuing large numbers of *explicit* ratings each. Smaller-sized, decentralized and open communities are typical for the Web 2.0. Here, ratings are mainly derived *implicitly* from user behavior and interaction patterns. However, these communities poorly qualify for blessings provided by recommender systems.

In this chapter, we explore an approach that intends to alleviate the information sparsity issue by exploiting *product classification taxonomies* as powerful background knowledge. Semantic product classification corpora for diverse fields are becoming increasingly popular, facilitating smooth interaction across company boundaries and fostering meaningful information exchange. For instance, the United Nations Standard Products and Services Classification (UNSPSC) contains over 11,000 codes [Obrst et al, 2003]. The taxonomies that Amazon.com (*http://www.amazon.com*) provides feature even more abundant, hierarchically arranged background knowledge: the book classification taxonomy alone comprises 13,500 topics, its pendant for categorizing movies and DVDs has about 16,400 concepts. Moreover, all products available on Amazon.com bear several descriptive terms referring to these taxonomies, thus making product descriptions machine-readable.

While the Amazon.com and the UNSPSC taxonomies have been crafted in a *controlled* environment by a small number of editors, the Web 2.0 has spawned projects that aim for much larger taxonomies, such as the DMOZ project, which intends to

C.-N. Ziegler: *Social Web Artifacts for Boosting Recommenders*, SCI 487, pp. 23–45.
DOI: 10.1007/978-3-319-00527-0_3   © Springer International Publishing Switzerland 2013

classify the entire Web into categories. This project alone is supported by more than 69,000 editors and has generated more than 590,000 category nodes categorizing more than 5.1 million sites. Taxonomies crafted by joint efforts of volunteers are thus a typical manifestation of the Web 2.0, which is why we want to make use of them for improving the quality of recommender systems.

Our novel taxonomy-based similarity metric, making inferences from hierarchical relationships between classification topics, represents the core of our hybrid filtering framework to compensate for sparsity. Quality recommendations become feasible in communities suffering from low information density, too.

We collected and crawled data from the very sparse All Consuming book readers' community (*http://www.allconsuming.net*) and conducted various experiments indicating that the taxonomy-driven method significantly outperforms benchmark systems. Repeating the offline evaluations for the dense MovieLens dataset, our approach's performance gains shrink considerably, but still exceed benchmark scores.

3.2   Related Work

Many attempts have been made to overcome the sparsity issue. Sarwar et al [2000b] propose singular value decomposition (SVD) as an efficient means to reduce the dimensionality of product-user matrices in collaborative filtering. Results reported have been mixed. Personal filtering agents [Good et al, 1999], surrogates for human users, represent another approach and have been shown to slightly improve results when deployed into sparse, human-only communities. Srikumar and Bhasker [2004] combine association rule mining and user-based CF to cope with sparsity.

The idea of using taxonomies for information filtering has been explored before, the most prominent example being directory-based browsing of information mines, e.g., Yahoo (*http://www.yahoo.com*), Google Directory (*http://www.google.com*), and ACM Computing Reviews (*http://www.reviews.com*). Moreover, Sollenborn and Funk [2002] propose *category*-based filtering, similar to the approach pursued by [Baudisch, 2001]. Pretschner and Gauch [1999] personalize Web search by using ontologies that represent user interests for profiling.

However, these taxonomy-based approaches do not exploit semantic "is-a" relationships between topics for profiling. Middleton et al [2001, 2002] recommend research papers, using ontologies to inductively learn topics that users are particularly interested in. Knowing a user's most liked topics then allows efficient product set filtering, weeding out those research papers that do not fall into these favorite topics. In contrast to our own technique proposed, Middleton uses clustering techniques for categorization and does not make use of human-created, large-scale product classification taxonomies.

3.3   Approach Outline

Following the "collaboration via content" paradigm [Pazzani, 1999], our approach computes content-based user profiles which are then used to discover like-minded peers. Once the active agent's neighborhood of most similar peers has been formed, the recommender focuses on products rated by those neighbors and generates top-N recommendation lists. The rank assigned to a product b depends on two factors. First, the similarity weight of neighbors voting for b, and, second, b's content description with respect to the active user's interest profile. Hence the hybrid nature of our approach.

3.3.1 Information Model

Before delving into algorithmic details, we introduce the formal information model, which can be tied easily to arbitrary application domains. Note that the model at hand also serves as foundation for subsequent chapters.

- **Agents** $A = \{a_1, a_2, \ldots, a_n\}$. All community members are elements of A. Possible identifiers are globally unique names, URIs, etc.
- **Product set** $B = \{b_1, b_2, \ldots, b_m\}$. All domain-relevant products are stored in set B. Unique identifiers either refer to proprietary product codes from an online store, such as Amazon.com's ASINs, or represent globally accepted standard codes, e.g., ISBNs.
- **User ratings** $R_1, R_2, \ldots, R_n$. Every agent a_i is assigned a set $R_i \subseteq B$ which contains his *implicit* product ratings. Implicit ratings, such as purchase data, product mentions, etc., are far more common in electronic commerce systems and online communities than explicit ratings [Nichols, 1998].
- **Taxonomy** C **over set** $D = \{d_1, d_2, \ldots, d_l\}$. Set D contains categories for product classification. Each category $d_e \in D$ represents one specific topic that products $b_k \in B$ may fall into. Topics express broad or narrow categories. The partial taxonomic order $C : D \to 2^D$ retrieves all immediate sub-categories $C(d_e) \subseteq D$ for topics $d_e \in D$. Hereby, we require that $C(d_e) \cap C(d_h) = \emptyset$ holds for all $d_e, d_h \in D, e \neq h$, and hence impose tree-like structuring, similar to single-inheritance class hierarchies known from object-oriented languages. Leaf topics d_e are topics with zero outdegree, formally $C(d_e) = \bot$, i.e., most specific categories. Furthermore, taxonomy C has exactly one top element, $\top$, which represents the most general topic and has zero indegree.
- **Descriptor assignment function** $f : B \to 2^D$. Function f assigns a set $D_k \subseteq D$ of product topics to every product $b_k \in B$. Note that products may possess *several* descriptors, for classification into one single category may be too imprecise.

3.3.2 Taxonomy-Driven Profile Generation

Collaborative filtering techniques represent user profiles by vectors $\mathbf{v_i} \in [-1, +1]^{|B|}$, where $v_{i,k}$ indicates the user's *rating* for product $b_k \in B$. Similarity between agents a_i and a_j is computed by applying Pearson correlation or cosine similarity to their respective profile vectors (see Section 2.3.2). Clearly, for very large $|B|$ and comparatively small $|A|$, this representation fails, owing to insufficient overlap of rating vectors.

We propose another, more informed approach which does not represent users by their respective *product*-rating vectors of dimensionality $|B|$, but by vectors of interest scores assigned to *topics* taken from taxonomy C over product categories $d \in D$.

User profile vectors are thus made up of $|D|$ entries, which corresponds to the number of distinct classification topics. Moreover, making use of profile vectors representing interest in *topics* rather than product *instances*, we can exploit the hierarchical structure of taxonomy C in order to generate overlap and render the similarity computation more meaningful: for every topic $d_{k_e} \in f(b_k)$ of products b_k that agent a_i has implicitly rated, we also infer an interest score for all *super*-topics of d_{k_e} in user a_i's profile vector. However, score assigned to super-topics decays with increasing distance from leaf node d_{k_e}. We furthermore normalize profile vectors with respect to the amount of score assigned, according the arbitrarily fixed overall score s.

Hence, suppose that $\mathbf{v_i} = (v_{i,1}, v_{i,2}, \ldots, v_{i,|D|})^T$ represents the profile vector for user a_i, where $v_{i,k}$ gives the score for topic $d_k \in D$. Then we require the following equation to hold:

$$\forall a_i \in A : \sum_{k=1}^{|D|} v_{i,k} = s \qquad (3.1)$$

By virtue of agent-wise normalization for a_i's profile, the score for each product $b_k \in R_i$ amounts to $s / |R_i|$, inversely proportional to the number of distinct products that a_i has rated. Likewise, for each topic descriptor $d_{k_e} \in f(b_k)$ categorizing product b_k, we accord topic score $\mathrm{sc}(d_{k_e}) = s / (|R_i| \cdot |f(b_k)|)$. Hence, the topic score for b_k is distributed evenly among its topic descriptors.

Let $(p_0, p_1, \ldots, p_q)$ denote the path from top element $p_0 = \top$ to descendant $p_q = d_{k_e}$ within the tree-structured taxonomy C for some given $d_{k_e} \in f(b_k)$. Then topic descriptor d_{k_e} has q super-topics. Score normalization and inference of fractional interest for super-topics imply that descriptor topic score $\mathrm{sc}(d_{k_e})$ may *not* become *fully* assigned to d_{k_e}, but in part to all its ancestors $p_{q-1}, \ldots p_0$, likewise. We therefore introduce another score function $\mathrm{sco}(p_m)$ that represents the eventual assignment of score to topics p_m along the taxonomy path leading from $p_q = d_{k_e}$ to $p_0 = \top$:

$$\sum_{m=0}^{q} \mathrm{sco}(p_m) = \mathrm{sc}(d_{k_e}) \qquad (3.2)$$

In addition, based on results obtained from research on semantic distance in taxonomies (e.g., see [Budanitsky and Hirst, 2000] and [Resnik, 1999]), we require that interest score $\mathrm{sco}(p_m)$ accorded to p_m, which is super-topic to p_{m+1}, depends on the number of siblings, denoted $\mathrm{sib}(p_{m+1})$, of p_{m+1}: the fewer siblings p_{m+1} possesses, the more interest score is accorded to its super-topic node p_m:

$$\mathrm{sco}(p_m) = \kappa \cdot \frac{\mathrm{sco}(p_{m+1})}{\mathrm{sib}(p_{m+1}) + 1} \qquad (3.3)$$

We assume that sub-topics have *equal shares* in their super-topic within taxonomy C. Clearly, this assumption may imply several issues and raise concerns, e.g., when certain sub-taxonomies are considerably denser than others [Resnik, 1995, 1999].

Propagation factor κ permits to fine-tune the profile generation process, depending on the underlying taxonomy's depth and granularity. For instance, we apply $\kappa = 0.75$ for Amazon.com's book taxonomy.

Equations 3.2 and 3.3 describe conditions which have to hold for the computation of leaf node p_q's profile score $\text{sco}(p_q)$ and the computation of scores for its taxonomy ancestors p_k, where $k \in \{0, 1, \ldots, q-1\}$. We hence derive the following recursive definition for $\text{sco}(p_q)$:

$$\text{sco}(p_q) := \kappa \cdot \frac{\text{sc}(d_{k_e})}{g_q},\qquad(3.4)$$

where

$$g_0 := 1, \; g_1 := 1 + \frac{1}{\text{sib}(p_q) + 1},$$

and $\forall n \in \{2, \ldots, q\}$

$$g_n := g_{n-1} + (g_{n-1} - g_{n-2}) \cdot \frac{1}{\text{sib}(p_{q-n+1}) + 1}$$

Computed scores $\text{sco}(p_m)$ are used to build a profile vector $\mathbf{v_i}$ for user a_i, adding scores for topics in $\mathbf{v_i}$. The procedure is repeated for every product $b_k \in R_i$ and every $d_{k_e} \in f(b_k)$.

Example 1 (Profile computation). Suppose taxonomy C as depicted in Figure 3.1, and propagation factor $\kappa = 1$. Let a_i have implicitly rated four books, namely Matrix Analysis, Fermat's Enigma, Snow Crash, and Neuromancer. For Matrix Analysis, five topic descriptors are given, one of them pointing to leaf topic ALGEBRA within our small taxonomy.

Suppose that $s = 1000$ defines the overall accorded profile score. Then the score assigned to descriptor ALGEBRA amounts to $s / (4 \cdot 5) = 50$. Ancestors of leaf ALGEBRA are PURE, MATHEMATICS, SCIENCE, and top element BOOKS. Therefore, score 50 must be distributed among these topics according to Equation 3.2 and 3.3. The application of Equation 3.4 yields score 29.091 for topic ALGEBRA. Likewise, applying Equation 3.3, we get 14.545 for topic PURE, 4.848 for MATHEMATICS, 1.212 for SCIENCE, and 0.303 for top element BOOKS. These values are then used to build profile vector $\mathbf{v_i}$ for a_i.

3.3.3 Neighborhood Formation

Taxonomy-driven profile generation computes flat profile vectors $\mathbf{v_i} \in [0, s]^{|D|}$ for agents a_i, assigning score values between 0 and maximum score s to every topic d from the set of product categories D. In order to generate neighborhoods of like-minded peers for the active user a_i, a proximity measure is required.

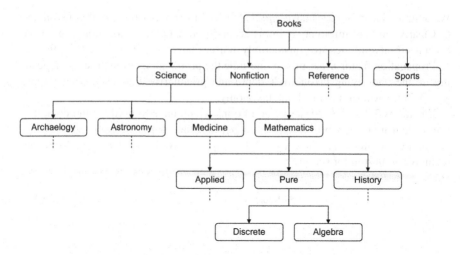

Fig. 3.1 Fragment from the Amazon.com book taxonomy

3.3.3.1 Measuring Proximity

Pearson's correlation coefficient and cosine similarity count among the most prominent correlation measures for CF (see Section 2.3.2.1). For our taxonomy-driven method, we opted for Pearson correlation, which Herlocker et al [2002] have found to perform better on collaborative filtering than cosine similarity.

Clearly, people who have implicitly rated many products in common also have high similarity. For generic collaborative filtering approaches, the proposition's inversion also holds, i.e., people who have *not* rated many products in common have *low* similarity.

On the other hand, applying taxonomy-driven profile generation, high similarity values can be derived even for pairs of agents that have *little* or even *no* products in common. Clearly, the measure's quality substantially depends on the taxonomy's design and level of nesting. According to our scheme, the more score two profiles v_i and v_j have accumulated in same branches, the higher their measured similarity.

Example 2 (Interest correlation). Suppose the active user a_i has rated only one single book b_m, bearing exactly one topic descriptor that classifies b_m into AL-GEBRA. User a_j has read a different book b_n whose topic descriptors point to diverse leaf nodes[1] of HISTORY, denoting history of mathematics. Then $c(a_i, a_j)$ will still be reasonably high, for both profiles have significant overlap in categories MATHEMATICS and SCIENCE.

Negative correlation occurs when users have *completely diverging interests*. For instance, in our information base mined from All Consuming, we had one user reading books mainly from the genres of Science Fiction, Fantasy, and Artificial Intelligence. The person in question was negatively correlated to another one reading books about American History, Politics, and Conspiracy Theories.

---

[1] Leaf nodes of HISTORY are not shown in Figure 3.1.

3.3.3.2 Selecting Neighbors

Having computed proximity weights $c(a_i, a_j)$ for the active user a_i and agents $a_j \in A \setminus \{a_i\}$, neighborhood formation takes place. Agent a_i's neighborhood, denoted by prox(a_i), contains a_i's most similar peers for use in computing recommendation lists.

Herlocker et al [1999] name two techniques for neighborhood selection, namely correlation-thresholding and best-M-neighbors. Correlation-thresholding puts users a_j with similarities $c(a_i, a_j)$ above some given threshold t into prox(a_i), whereas best-M-neighbors picks the M best correlates for a_i's neighborhood.

We opted for best-M-neighbors, since correlation-thresholding implies diverse unwanted effects when sparsity is high [Herlocker et al, 1999].

3.3.4 Recommendation Generation

Candidate products for a_i's personalized recommendation list are taken from his neighborhood's implicit ratings, avoiding products that a_i already knows:

$$B_i = \bigcup \{R_j \mid a_j \in \text{prox}(a_i)\} \setminus R_i \qquad (3.5)$$

Candidates $b_k \in B_i$ are then weighted according to their *relevance* for a_i. The relevance of products $b_k \in B_i$ for a_i, denoted $w_i(b_k)$, depends on various factors. Most important, however, are two aspects:

- **User proximity.** Similarity measures $c(a_i, a_j)$ of all those agents a_j that "recommend" product b_k to the active agent a_i are of special concern. The closer these agents to a_i's interest profile, the higher the relevance of b_k for a_i. We borrowed the latter intuition from common collaborative filtering techniques (see Section 1.1.1).
- **Product proximity.** Second, measures $c_b(a_i, b_k)$ of product b_k's closeness with respect to a_i's interest profile are likewise important. The purely content-based metric supplements the overall recommendation generation process with more fine-grained filtering facilities: mind that even highly correlating agents may appreciate items beyond the active user's specific interests. Otherwise, these agents would have *identical* interest profiles, not just similar ones.

  The computation of $c_b(a_i, b_k)$ derives from the user similarity computation scheme. For this purpose, we create a *dummy user* a_θ with $R_\theta = \{b_k\}$ and define $c_b(a_i, b_k) := c(a_i, a_\theta)$.

The relevance $w_i(b_k)$ of product b_k for the active user a_i is then defined by the following formula:

$$w_i(b_k) = \frac{q \cdot c_b(a_i, b_k) \cdot \sum_{a_j \in A_i(b_k)} c(a_i, a_j)}{|A_i(b_k)| + \Upsilon_R}, \qquad (3.6)$$

where
$$A_i(b_k) = \{a_j \in \text{prox}(a_i) \mid b_k \in R_j\}$$
and
$$q = (1.0 + |f(b_k)| \cdot \Gamma_T)$$

Variables Υ_R and Γ_T represent *fine-tuning parameters* that allow for customizing the recommendation process. Parameter Υ_R penalizes products occurring infrequently in rating profiles of neighbors $a_j \in \text{prox}(a_i)$. Hence, large Υ_R makes popular items acquire higher relevance weight, which may be suitable for users wishing to be recommended well-approved and common products instead of rarities. On the other hand, low Υ_R treats popular and uncommon, new products in exactly the same manner, helping to alleviate the *latency problem* [Sollenborn and Funk, 2002]. For experimental analysis, we tried values between 0 and 2.5.

Parameter Γ_T rewards products b_k that carry *many* content descriptors, i.e., have large $|f(b_k)|$. Variable Γ_T proves useful because profile score normalization and super-topic score inference may penalize products b_k containing several, detailed descriptors $d \in f(b_k)$, and favor products having few, more general topic descriptors. Reward through Γ_T is assigned linearly by virtue of $(|f(b_k)| \cdot \Gamma_T)$. Consider that the implementation of exponential decay appears likewise reasonable, therefore reducing Υ_R's gain in influence when $|f(b_k)|$ becomes larger. However, we have not tried this extension.

Eventually, product relevance weights $w_i(b_k)$ computed for every $b_k \in B_i$ are used to produce the active user a_i's recommendation list. The injective function P_{w_i} : $\{1,2,\ldots,|B_i|\} \to B$ reflects recommendation ranking according to w_i in *descending* order. For top-N recommendations, all entries $P_{w_i}(k), k > N$ are discarded.

3.3.5 Topic Diversification

A technique we call *topic diversification* constitutes another cornerstone contribution of this chapter. The latter method represents an *optional* procedure to supplement recommendation generation and to enhance the computed list's utility for agent a_i.

The idea underlying topic diversification refers to providing an active user a_i with recommendations from *all* major topics that a_i has declared specific interest in. The following example intends to motivate our method:

Example 3 (Topic overfitting). Suppose that a_i's profile contains books from Medieval Romance, Industrial Design, and Travel. Suppose Medieval Romance has a 60% share in a_i's profile, Industrial Design and Travel have 20% each. Consequently, Medieval Romance's predominance will result in most recommendations originating from this super-category, giving way for Industrial Design and Travel not before all books from like-minded neighbors fitting well into the Medieval Romance shape have been inserted into a_i's recommendations.

We observe the above issue with many recommender systems using content-based and hybrid filtering techniques. For purely collaborative approaches, recommendation diversification according to the active user a_i's topics of interest becomes even less controllable. Remember that collaborative filtering does *not* consider the content of products but only ratings assigned.

3.3.5.1 Recommendation Dependency

In order to implement topic diversification, we assume that recommended products $P_{w_i}(o)$ and $P_{w_i}(p)$, along with their content descriptions, effectively *do* exert an impact on each other, which is commonly ignored by existing approaches: usually, only relevance weight ordering $o < p \Rightarrow w_i(P_{w_i}(o)) \geq w_i(P_{w_i}(p))$ must hold for recommendation list items.

To our best knowledge, Brafman et al [2003] are the only researchers assuming dependencies between recommendations. Their approach considers recommendation generation as inherently *sequential* and uses *Markov decision processes* (MDP) in order to model interdependencies between recommendations. However, apart from the idea of dependence between items $P_{w_i}(o)$, $P_{w_i}(p)$, Brafman's focus significantly differs from our own, emphasizing the economic *utility* of recommendations with respect to past and future purchases.

In case of our topic diversification technique, recommendation interdependence signifies that an item b's current *dissimilarity* with respect to preceding recommendations plays an important role and may influence the "new" ranking order. Algorithm 3.1 depicts the entire procedure, a brief textual sketch is given in the next few paragraphs.

procedure diversify $(P_{w_i} : \{1,\ldots,|B_i|\} \to B, \Theta_F \in [0,1])$ {

$\quad B_i \leftarrow \{P_{w_i}(k) \mid k \in [1, x \cdot N]\}; P_{w_i*}(1) \leftarrow P_{w_i}(1);$

$\quad$ **for** $z \leftarrow 2$ **to** N **do**

$\quad\quad$ set $B_i' \leftarrow B_i \setminus \{P_{w_i*}(k) \mid k \in [1, z[\};$

$\quad\quad \forall b \in B'$: compute $c^*(b, \{P_{w_i*}(k) \mid k \in [1, z[\});$

$\quad\quad$ compute $P_{c^*} : \{1, 2, \ldots, |B_i'|\} \to B_i'$ using $c^*;$

$\quad\quad$ **for all** $b \in B_i'$ **do**

$\quad\quad\quad P_{c^*}^{rev^{-1}}(b) \leftarrow |B_i'| - P_{c^*}^{-1}(b);$

$\quad\quad\quad w_i^*(b) \leftarrow P_{w_i}^{-1}(b) \cdot (1 - \Theta_F) + P_{c^*}^{rev^{-1}}(b) \cdot \Theta_F;$

$\quad\quad$ **end do**

$\quad\quad P_{w_i*}(z) \leftarrow \min\{w_i^*(b) \mid b \in B_i'\};$

$\quad$ **end do**

$\quad$ **return** $P_{w_i*};$

}

Alg. 3.1 Sequential topic diversification

3.3.5.2 Topic Diversification Algorithm

Function P_{w_i*} denotes the new recommendation list, resulting from the application of topic diversification. For every list entry $z \in [2,N]$, we collect those products b from the candidate products set B_i that do not occur in positions $o < z$ in P_{w_i*} and compute their similarity with set $\{P_{w_i*}(k) \mid k \in [1,z[\}$, which contains all new recommendations preceding rank z. We hereby compute the mentioned similarity measure, denoted $c^*(b)$, by applying our scheme for taxonomy-driven profile generation and proximity measuring presented in Section 3.3.2 and 3.3.3.1.

Sorting all products b according to $c^*(b)$ in reverse order, we obtain the *dissimilarity rank* $P_{c^*}^{rev}$. This rank is then merged with the original recommendation rank P_{w_i} according to diversification factor Θ_F, yielding final rank P_{w_i*}. Factor Θ_F defines the *impact* that dissimilarity rank $P_{c^*}^{rev}$ exerts on the eventual overall output. Large $\Theta_F \in [0.5,1]$ favors diversification over a_i's original relevance order, while low $\Theta_F \in [0,0.5[$ produces recommendation lists closer to the original rank P_{w_i}. For experimental analysis, we used diversification factors $\Theta_F \in [0,0.9]$.

Note that the ordered input lists P_{w_i} must be *considerably larger* than the eventual top-N list. Algorithm 3.1 uses constant x for that purpose. In our later experiments, we assumed $x = 4$, hence using top-80 input lists for final top-20 recommendations.

3.3.5.3 Osmotic Pressure Analogy

The effect of dissimilarity bears traits similar to that of osmotic pressure and selective permeability known from molecular biology (e.g., see Tombs [1997]): steady insertion of products b_o, taken from one specific area of interest d_o, into the recommendation list equates to the passing of molecules from one specific substance through the cell membrane into cytoplasm. With increasing concentration of d_o, owing to the membrane's selective permeability, the pressure for molecules b from other substances d rises. When pressure gets sufficiently high for one given topic d_p, its best products b_p may "diffuse" into the recommendation list, even though the original rank $P_{w_i}^{-1}(b_p)$ might be inferior to the rank of candidates from the prevailing domain d_o. Consequently, pressure for d_p decreases, paving the way for another domain for which pressure peaks.

Topic diversification hence resembles the membrane's selective permeability, which allows cells to maintain their internal composition of substances at required levels.

3.4 Offline Experiments and Evaluation

The following sections present empirical results that were obtained from evaluating our approach. Core engine parts of our system, along with most other software tools for data extraction and screen scraping, were implemented in Java, small portions in Perl. Remote access via Web interfaces was rendered feasible through PHP frontends.

Besides our taxonomy-driven approach, we also implemented three other recommender algorithms for comparison.

3.4.1 Data Acquisition

Experimentation, parameterization, and fine-tuning were conducted on "real-world" data, obtained from All Consuming (*http://www.allconsuming.net*), an open community addressing people interested in reading books. We extracted additional, taxonomic background knowledge, along with content descriptions of those books, from Amazon.com. Crawling was started on January 12 and finished on January 16, 2004.

The entire dataset comprises 2,783 users, representing either "real", registered members of All Consuming, or personal weblogs collected by the community's spiders, and 14,591 ratings addressing 9,237 diverse book titles. All ratings are implicit, i.e., non-quantifiable with respect to the extent of appreciation of the respective books. On average, users provided 5.24 book ratings.

After the application of various data cleansing procedures and duplicate removal, Amazon.com's tree-structured book classification taxonomy contained 13,525 distinct concepts. Our crawling tools collected 27,202 topic descriptors from Amazon.com, relating 8,641 books to the latter concept lattice. Consequently, for 596 of those 9,237 books mentioned by All Consuming's users, no content information could be obtained from Amazon.com, signifying only 6.45% rejects. We eliminated these books from our dataset. On average, 3.15 topic descriptors were found for books available on Amazon.com, thus making content descriptions sufficiently explicit and reliable for profile generation.

To make the analysis data obtained from our performance trials more accurate, we relied upon an external Web-service[2] to spot ISBNs referring to the same book, but different editions, e.g., hardcover and paperback. Those ISBNs were then mapped to one single representative ISBN.

3.4.2 Evaluation Framework

Since our taxonomy-driven recommender system operates on *binary* preference input, i.e., implicit rather than explicit ratings, predictive accuracy metrics (see Section 2.4.1.1) are not suitable for evaluation. We hence opted for decision-support accuracy metrics (see Section 2.4.1.2), namely precision, recall, and Breese score.

3.4.2.1 Benchmark Systems

Besides our own, taxonomy-driven proposal, we implemented three other recommendation algorithms: one "naïve", random-based system offering no personalization at all and therefore defining the bottom line, one purely collaborative approach,

---

[2] See http://www.oclc.org/research/projects/xisbn/

typically used for evaluations, and one hybrid method, exploiting content information provided by our dataset.

Bottom Line Definition

For any given user a_i, the naïve system randomly selects an item $b \in B \setminus R_i$ for a_i's top-N list $P_i : \{1, 2, \ldots, N\} \to B$. Clearly, as is the case for every other presented approach, products may not occur more than once in the recommendation list, i.e., $\forall o, p \in \{1, 2, \ldots, N\}, o \neq p : P_i(o) \neq P_i(p)$ holds.

The random-based approach shows results obtained when no filtering takes place, representing the base case that "non-naïve" algorithms are supposed to surpass.

Collaborative Filtering Algorithm

The common user-based CF algorithm, featuring extensions proposed by Herlocker et al [2002], traditionally serves as benchmark when evaluating recommender systems operating on *explicit* preferences. Sarwar et al [2000b] propose an adaptation specifically geared towards implicit ratings, known as "most frequent items". We used that latter system as CF benchmark, computing relevance weights $w_i(b_k)$ for books b_k from a_i's candidates set B_i according to the following scheme:

$$w_i(b_k) = \sum_{a_j \in A_i(b_k)} c(a_i, a_j) \tag{3.7}$$

Set $A_i(b_k) \subseteq \text{prox}(a_i)$ contains all neighbors of a_i who have implicitly rated b_k.

We measure user similarity $c(a_i, a_j)$ according to Pearson correlation (see Section 2.3.2.1). Profile vectors $\mathbf{v_i}, \mathbf{v_j}$ for agents a_i, a_j, respectively, represent implicit ratings for every product $b_k \in B$, hence $\mathbf{v_i}, \mathbf{v_j} \in \{0, 1\}^{|B|}$.

Hybrid Recommender Approach

The third system exploits both collaborative and content-based filtering techniques, representing user profiles $\mathbf{v_i}$ through collections of descriptive terms, along with their frequency of occurrence.

Descriptive terms for books b_k correspond to topic descriptors $f(b_k)$, originally relating book content to taxonomy C over categories D. Consequently, profile vectors $\mathbf{v_i} \in \mathbb{N}^{|D|}$ for agents a_i take the following shape:

$$\forall d \in D : v_{i,d} = |\{b_k \in R_i \mid d \in f(b_k)\}| \tag{3.8}$$

Neighborhoods are formed by computing Pearson correlations between all pairs of content-driven profile vectors and selecting best-M neighbors. Relevance is then defined as below:

$$w_i(b_k) = \frac{c_b(a_i, b_k) \cdot \sum_{a_j \in A_i(b_k)} c(a_i, a_j)}{|A_i(b_k)|} \tag{3.9}$$

Mind that Equation 3.9 presents a *special case* of Equation 3.6, assuming $\Gamma_T = 0$ and $\Upsilon_R = 0$. Essentially, the depicted hybrid approach constitutes a simplistic adaptation of our taxonomy-driven system. Differences largely refer to the underlying algorithm's lack of super-topic score inference, one major cornerstone of our novel method, and the lack of parameterization.

3.4.2.2 Experiment Setup

The evaluation framework intends to compare the *utility* of recommendation lists generated by all four recommender systems, applying precision, recall, and Breese score (see Section 2.4.1.2). In order to obtain *global* metrics, we averaged the respective metric values for all evaluated users.

First, we selected all users a_i with more than five ratings and discarded those having fewer ratings, owing to the fact that reasonable recommendations are beyond feasibility for these cases.

For cross-validation, we applied 5-folding, effectively performing 80/20 splits of every user a_i's implicit ratings R_i into five pairs of *training sets* R_i^x and *test sets* T_i^x, where $T_i^x = R_i \setminus R_i^x$. Consequently, we computed five complete recommendation lists for every a_i, i.e., one list for each $R_i^x, x \in \{1, \ldots, 5\}$.

3.4.2.3 Parameterization

We defined $|\text{prox}(a_i)| = 20$, i.e., requiring neighborhoods to contain exactly 20 peers, and we provided top-20 recommendations for each active user a_i's training set R_i^x. Similarities between profiles, based upon R_i^x and the original ratings R_j of all other agents a_j, were computed anew for each training set R_i^x of a_i.

For performance analysis, we parameterized our taxonomy-driven recommender system's profile generation process by assuming propagation factor $\kappa = 0.75$, which encourages super-topic score inference. We opted for $\kappa < 1$ since Amazon.com's book taxonomy is deeply-nested and topics tend to have numerous siblings, which makes it rather difficult for topic score to reach higher levels.

For recommendation generation, we adopted parameter $\Upsilon_R = 0.25$, i.e., books occurring *infrequently* in ratings issued by the active user's neighbors were therefore not overly penalized. Generous reward was accorded for books b_k bearing *detailed* content descriptions, i.e., having large $|f(b_k)|$, by assuming $\Gamma_T = 0.1$. Hence, a 10% bonus was granted for every additional topic descriptor. For topic diversification, we adopted $\Theta_F = 0.33$.

No parameterizations were required for the random-based, purely collaborative, and hybrid approaches.

3.4.2.4 Result Analysis

We measured performance by computing precision, recall, and Breese score, assuming half-life $\alpha = 5$ and $\alpha = 10$, for all four recommenders and all combinations of training and test sets. Results are displayed in Figure 3.2 and 3.3.

For each indicated chart, the horizontal axis expresses the *minimum number* of ratings that users were required to have issued so they were considered for recommendation generation and evaluation. Note that larger x-coordinates hence imply that *fewer* agents were considered for computing the respective data points.

Results obtained seem to prove our hypothesis that taxonomy-driven recommendation generation outperforms common approaches when dealing with sparse product rating information: all four metrics position our novel approach *significantly* above its purely collaborative and hybrid counterparts.

We observe one considerable cusp common to all four charts and particularly pronounced for the taxonomy-based curves. The sudden drop happens when users bearing exactly 36 implicit ratings become discarded. On average, for the taxonomy-driven recommendation generation, these agents have high ranks with respect to all four metrics applied. Removal thus temporarily lowers the curves.

More detailed, metric-specific analysis follows in subsequent paragraphs.

Precision

Surprisingly, precision increases even for the random recommender when ignoring users with few ratings. The reason for this phenomenon lies in the nature of the precision metric: for users a_i with test sets T_i^x smaller than the number $|P_i^x|$ of recommendations received, i.e., $|T_i^x| < 20$, there is *no possibility* of achieving 100% precision.

Analysis of unweighted precision, given in Figure 3.2(b), shows that the gap between our taxonomy-driven approach and its collaborative and hybrid pendants becomes even larger when users are required to have rated many books. Agents with small numbers of ratings tend to perturb prediction accuracy as no proper "guidance" for neighborhood selection and interest definition can be provided.

Differences between the collaborative and the hybrid method are less significant and rather marginal. However, the first steadily outperforms the former when making recommendations for agents with numerous ratings.

Unweighted and Weighted Recall

Unweighted recall, shown in Figure 3.2(b), presents a slightly different scenario: even though the performance gap between the taxonomy-driven recommender and both other, non-naïve methods still persists, it does not become larger for increasing x. Collaborative filtering, slightly inferior to its hybrid pendant at first, overtakes the latter when considering agents with numerous ratings only. Similar observations have been made by Pazzani [1999].

Figure 3.3 allows more fine-grained analysis with respect to the accuracy of rankings. Remember that unweighted recall is equivalent to Breese score when assuming half-life $\alpha = \infty$ (see Section 2.4.1.2). While pure collaborative filtering shows

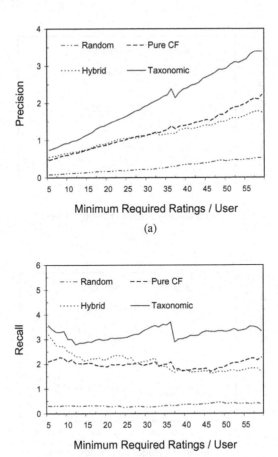

Fig. 3.2 Unweighted precision (a) and recall (b) metrics

largely insensitive to decreasing α, hybrid and taxonomy-driven recommenders do not. Assuming $\alpha = 10$, the first derivative of the latter two systems improves over their corresponding recall curves for increasing x-coordinates. This notable development becomes even more pronounced when further decreasing half-life to $\alpha = 5$.

Consequently, in case of content-exploiting methods, relevant products $b \in \Im P_i^x \cap T_i^x$ have the tendency to appear "earlier" in recommendation lists P_i^x, i.e., have comparatively small distance from the top rank. On the other hand, for collaborative filtering, relevant products seem to be more uniformly distributed among top-20 ranks.

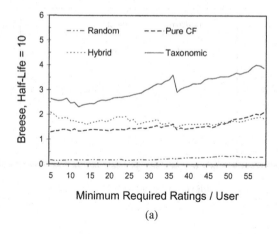

(a)

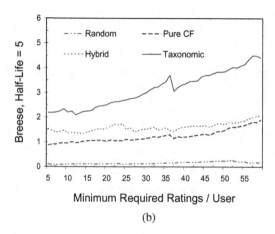

(b)

Fig. 3.3 Weighted recall, using half-life $\alpha = 10$ (a) and $\alpha = 5$ (b)

3.5 Deployment and Online Study

On February 9, 2004, we deployed our taxonomy-driven recommender system into the All Consuming community[3], providing personalized recommendations for registered users based upon their book rating profile. Access facilities are offered through diverse PHP scripts that query an RDBMS containing rating profiles, neighborhood information, and precomputed recommendations, likewise.

Besides our taxonomy-driven approach, we also implanted both other non-naïve approaches documented before into All Consuming. Registered users could hence

---

[3] The computed recommendation lists can be reached through All Consuming's *News*-section, see http://cgi.allconsuming.net/news.html

access *three distinct lists* of top-20 recommendations, customized according to their personal rating profile. We utilized the depicted system setup to conduct online performance comparisons, going beyond offline statistical measures.

3.5.1 Online Study Setup

For the online evaluation, we demanded All Consuming members to rate all recommendations provided on a 5-point likert scale, ranging from −2 to +2. Hereby, raters were advised to give *maximum score* for recommended books they had *already read*, but not indicated in their reading profile. Moreover, users were given the opportunity to return an *overall* satisfaction verdict for each recommendation list. The additional rating served as an instrument to also reflect the make-up and quality of list composition. Consequently, members could provide 63 rating statements each.

3.5.2 Result Analysis

54 All Consuming members, not affiliated with our department and university, volunteered to participate in our online study by December 3, 2004. They provided 2,164 ratings about recommendations they were offered, and 131 additional, overall list quality statements. Since not every user rated all 60 books recommended by our three diverse systems, we assumed neutral votes for recommended books not rated. Furthermore, in order not to bias users towards our taxonomy-driven approach, we assigned letters A, B, C to recommendation lists, not revealing any information about the algorithm operating behind the scenes.

While 50 users rated one or more recommendations computed according to the purely collaborative method, named A, 46 did so for the taxonomy-driven approach, labelled B, and 42 for the simplistic hybrid algorithm. In a first experiment, depicted in Figure 3.4(a), we compared the overall recommendation list statements and average ratings of personalized top-20 recommendations for each rater and each recommender system. Results were averaged over all participating users. In both cases, the taxonomy-driven system performed best and the purely collaborative worst.

Second, we counted all those raters perceiving one specific system as best. Again, the comparison was based upon the overall statements and average recommendation ratings, likewise. In order to guarantee fairness, we discarded users not having rated all three systems for each metric. Figure 3.4(b) shows that the taxonomy-driven method outperforms both other recommendation techniques.

Eventually, we may conclude that results obtained from the online analysis back our offline evaluation results. In both cases, the taxonomy-driven method has been shown to outperform benchmark systems for the sparse All Consuming dataset.

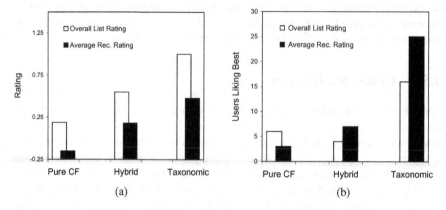

Fig. 3.4 Results obtained from the online study

3.6 Movie Data Analysis

The dataset we obtained from crawling the All Consuming community exhibits two properties we believe pivotal for the superiority of our technique over common benchmark methods:

- **Rating information sparseness.** Compared to the number of ratings, the number of unique ISBNs is relatively large. Moreover, most users have issued few ratings, these being implicit only. Hence, the probability of having product rating profiles with numerous overlapping products is low, implying bad performance scores for standard collaborative filtering approaches. On the other hand, taxonomy-driven profiling has been conceived to directly address these issues and to render sparse profile vectors dense.
- **Fine-grained domain classification.** Books address most aspects of our every-day life, e.g., education, business, science, entertainment, and so forth. Therefore, the construction of highly nested and detailed classification taxonomies becomes feasible. Moreover, owing to comparatively high *costs of consumption*[4], people *deliberately* consume products matching their specific interests only. Inspection of purchase and book reading histories clearly reveals these diverse interests and makes profile compositions easily discernable, which is essential for finding appropriate neighbors.

However, we would like to test our approach on domains where the two aforementioned assumptions do not hold anymore. We hence opted for the popular MovieLens dataset [Sarwar et al, 2001, 2000b], which contains *explicit* ratings about movies and has a very high density.

---

[4] Note that reading books takes much more time than watching DVDs or listening to CDs.

Movies bear intrinsic features that make them largely different from books. For instance, their cost of consumption tends to be much lower. Consequently, people are more inclined to experience products that may not perfectly match their profile of interest. We conjecture that such exploratory behavior makes interest profiles, inferred from implicit or explicit ratings, less concise and less accurate.

Moreover, movies are basically geared towards the entertainment sector only, not spanning other areas of life, e.g., science, business, and so forth. We believe both aspects disadvantageous for taxonomy-driven profiling.

3.6.1 Dataset Composition

The small MovieLens dataset contains 943 users who have issued $100,000$ explicit ratings on a 5-point likert scale, referring to $1,682$ movies. The average number of ratings per user hence amounts to 106.04, meaning that the average user has rated 6.31% of all ratable products. These numbers highly contrast All Consuming's figures, where the average user has rated 5.24 books and thus only 0.04% of the entire product set.

In order to make taxonomy-driven recommendations feasible, we crawled Amazon.com's movie taxonomy, extracting $16,481$ hierarchically arranged topics. This number clearly exceeds the book taxonomy's $13,525$ concepts. In addition, both lattices exhibit subtly different characteristics with respect to structure: the movie taxonomy's average distance from root to leafs amounts to 4.25, opposed to 5.05 for books. However, the average number of inner node siblings is higher for movies than for books, contrasting 18.53 with 16.65. Hence, we may conclude that the book taxonomy is deeper, though more condensed, than its movie pendant.

We were able to obtain taxonomic descriptions for 1519 of all 1682 movies on MovieLens from Amazon.com, collecting 9281 descriptors in total. On average, 5.52 topic descriptors were found for those movies for which content information could be provided. The remaining 163 movies were removed from the dataset, along with all 8668 ratings referring to them.

3.6.2 Offline Experiment Framework

We opted for roughly the same analysis setup as presented for the All Consuming offline evaluations. Since MovieLens features explicit ratings, we denote user a_i's ratings by function $r_i : B \to \{1, 2, \ldots, 5\}^\perp$ rather than $R_i \subseteq B$. We tailored our evaluation metrics and benchmark recommenders in order to account for explicit ratings.

3.6.2.1 Benchmark Systems

The parameters for the taxonomy-driven approach were slightly modified in order to optimize results. For topic diversification, we supposed $\Theta_F = 0.25$. Super-topic score inference was promoted by assuming $\kappa = 1.0$. Moreover, only movies b_k that had been assigned an excellent rating of 4 or 5, i.e., $b_k \in \{b \in B \mid r_i(b) \geq 4\}$, were considered for the generation of a_i's profile.

The random-based recommender was kept in order to mark the absolute bottom line.

Collaborative Filtering Algorithm

Instead of "most frequent items" [Sarwar et al, 2000b], we used the original Group-Lens collaborative recommender [Konstan et al, 1997; Resnick et al, 1994], which had been specifically designed with *explicit* ratings in mind (see Section 2.3.2.1). We extended the system by implementing modifications proposed by Herlocker et al [1999], i.e., significance weighting, deviation from mean, and best-M neighborhood formation, in order to render the recommender as competitive as possible. We found that the application of significance weighting, i.e., penalizing high correlation values based upon less than 50 products in common, increased the system's performance *substantially*.

Most Popular Products Recommender

Breese et al [1998] compare benchmarks against an efficient, though non-personalized recommender. The algorithm simply proposes overall top-N most rated products to the active user a_i. However, these products are required not to occur in a_i's training set R_i^x, i.e., $P_i^x \cap R_i^x = \emptyset$, implying that recommendation lists P_i^x, P_j^x for two different users a_i, a_j are not completely identical.

3.6.2.2   Setup and Metrics

Again, we applied 5-folding cross-validation and assumed neighborhoods of dimension $|\text{prox}(a_i)| = 20$ for all users a_i. In contrast to the All Consuming experimental setup, we provided top-10 recommendations instead of top-20.[5]

Moreover, the input test sets we provided to precision and recall slightly differed from the input provided in preceding experiments: in order to account for the fact that all ratings were *explicit*, i.e., that we actually *knew* if user a_i had liked product b_k experienced earlier, only test set products $b_k \in \{b \in T_i^x \mid r_i(b) \geq 4\}$ were counted as hits, i.e., those products that a_i had assigned *positive* ratings:

$$\text{Recall} = 100 \cdot \frac{|\Im P_i^x \cap \{b \in T_i^x \mid r_i(b) \geq 4\}|}{|\{b \in T_i^x \mid r_i(b) \geq 4\}|} \qquad (3.10)$$

Accordingly, the computation of precision with rating-constrained test set input looks as follows:

$$\text{Precision} = 100 \cdot \frac{|\Im P_i^x \cap \{b \in T_i^x \mid r_i(b) \geq 4\}|}{|\Im P_i^x|} \qquad (3.11)$$

---

[5] We found little variation in precision/recall scores when decreasing the recommendation list size from 20 to 10.

We also computed F1 scores (see Section 2.4.1.2), based upon the above-given versions of precision and recall.

3.6.2.3 Result Analysis

Precision and recall values considering the complete 943 users dataset were computed for all four recommender systems. The respective scores are given by Figure 3.5. One can see that the obtained metric values tend to be *higher* than their equivalents for the All Consuming community data. Apart from the random recommender[6], all algorithms achieved more than 10% recall and 12% precision. The reasons for these comparatively high scores lie primarily in the much larger density of MovieLens as opposed to All Consuming, indicated before in Section 3.6.1.

The taxonomy-driven approach, having an F1 metric score of 17.59%, outperforms the purely collaborative system as well as the non-personalized recommender for top-N most popular products. The latter method also shows inferior to the collaborative filter, made explicit by an F1 score of 11.34% versus 13.98%.

However, the relative performance gap between our taxonomy-driven approach and its benchmark recommenders is definitely more pronounced for the All Consuming book rating data than for MovieLens. Conjectures about possible reasons have already been mentioned in Section 3.6, counting domain-dependence and rating sparsity among the major driving forces.

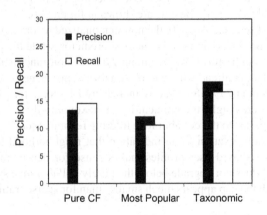

Fig. 3.5 Performance analysis for the complete MovieLens dataset

---

[6] The random recommender maintained precision/recall values far below 1% and is not displayed in Figure 3.5.

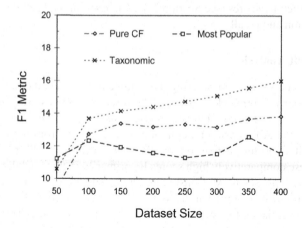

Fig. 3.6 MovieLens dataset size sensitivity

Dataset Size Sensitivity

In a second experiment, we tested the *sensitivity* of all presented non-naïve recommenders with respect to the numbers of users eligible for neighborhood formation. Neither the product set size nor the number of ratings per user were modified. We created 8 subsets of the MovieLens user base, selecting the first $x \cdot 50$ users from the complete set, $x \in \{1, 2, \ldots, 8\}$. Results are shown in Figure 3.6.

For the smallest set, i.e., $|A| = 50$, the non-personalized recommender for overall most appreciated products shows to be the best predictor, while the purely collaborative filtering system performs worst among the three non-random recommenders. However, for 100 users already, the two personalized approaches overtake the non-personalized system and exhibit steadily increasing F1 scores for increasing numbers of users x. Interestingly, the gradient for the taxonomy-driven method's curve is still slightly superior to the collaborative filtering recommender's.

We regard this observation as an indication that neighborhood formation relying upon taxonomy-based user profiles makes sense for denser rating data, too. The accuracy still does not degrade below the purely collaborative system's benchmark, even though the gap appears much smaller than for sparser rating information scenarios.

3.7 Conclusion

In this chapter, we presented a novel, hybrid approach to automated recommendation making, based upon large-scale product classification taxonomies which are readily available for diverse domains on the Internet. Cornerstones of our approach are the *generation of profiles* via inference of super-topic score and the *recommendation framework* itself.

Offline performance trials were conducted on "real-world" data in order to demonstrate our algorithm's superiority over less informed approaches when rating information sparseness prevails. Moreover, we conducted online studies, asking All Consuming community members to rate and compare diverse recommender systems. In addition to sparse book rating information, we tested our approach's performance when dealing with substantially different data, running benchmark comparisons on the well-known MovieLens dataset. Results suggested that taxonomy-driven recommending still performs better on denser data than competing systems. However, the performance gap becomes comparatively small and does no longer justify additional efforts for acquiring costly domain knowledge, which taxonomy-driven filtering substantially depends on.

Chapter 4
Topic Diversification Revisited

"Nothing is pleasant that is not spiced with variety."

– Francis Bacon (1561 – 1626)

4.1 Introduction

Chapter 3 has introduced topic diversification as an efficient means to avoid *topic overfitting* in our taxonomy-driven filtering approach. However, the topic diversification method can be applied to *any* recommender system that generates ordered top-N lists of recommendations, as long as taxonomic domain knowledge is available for the recommendation domain in question.

Winners Take All

The main reason for the a posteriori application of topic diversification to conventional recommender systems lies in the fact that most recommender algorithms are highly susceptible to *winners-take-all* behavior: soon as the user's profile bucket contains one subset of similar products that appears larger than any other similarity-based subset, the chances that *all* computed recommendations will derive from that cluster are high. The observation can be made for algorithms using content-based similarity measures, and techniques based on collaborative similarity metrics, e.g., item-based CF (see Section 2.3.2.2), likewise. For instance, many people complain that Amazon.com's (*http://www.amazon.com*) recommendations, computed according to the item-based CF scheme [Linden et al, 2003], appear too "similar" with respect to content. Hence, customers that have purchased many books written by Herrmann Hesse may happen to obtain recommendation lists where *all* top-5 entries contain books from that respective author only. When considering pure accuracy, all these recommendations seem excellent, since the active user clearly appreciates Hesse's novels. On the other hand, assuming that the active user has several interests other than Herrmann Hesse, e.g., historical novels in general and books about world travel, the recommended set of items appears poor, owing to its lack of diversity.

C.-N. Ziegler: *Social Web Artifacts for Boosting Recommenders*, SCI 487, pp. 47–59.
DOI: 10.1007/978-3-319-00527-0_4   © Springer International Publishing Switzerland 2013

Some researchers, e.g., Ali and van Stam [2004], have noticed the depicted issue, commonly known as the "portfolio effect", for other recommender systems before. However, to our best knowledge, no solutions have been proposed so far.

Reaching Beyond Accuracy

Topic diversification can solve the portfolio effect issue, balancing and diversifying personalized recommendation lists to reflect the user's *entire spectrum* of interests. However, when running offline evaluations based upon accuracy metrics such as precision, recall, and MAE (see Section 2.4.1), we may expect the performance of topic diversification-enhanced filters to show *inferior* to that of their respective non-diversified pendants. Hence, while believed beneficial for actual user satisfaction, we conjecture that topic diversification will prove detrimental to accuracy metrics.

For evaluation, we therefore pursue a twofold approach, conducting one large-scale online study that involves more than 2,100 human subjects, and offline analysis runs based on 361,349 ratings. Both evaluation methods feature the application of diverse degrees of diversification to the two most popular recommendation techniques, i.e., user-based and item-based CF (see Section 2.3.2). The bilateral evaluation approach renders the following types of result analysis possible:

- **Accuracy and diversity.** The application of precision and recall metrics to user-based and item-based CF with varying degrees of diversification, $\Theta_F \in [0.1, 0.9]$, exposes the *negative* impacts that topic diversification exerts on accuracy. Proposing the offline *intra-list similarity* measure, we are able to capture and quantify the diversity of top-N recommendation lists, with respect to one given similarity metric. Contrasting the measured accuracy and diversity, their overall *negative* correlation becomes revealed.
- **Topic diversification benefits and limitations.** The online study shows that users tend to appreciate diversified lists. For diversification factors $\Theta_F \in [0.3, 0.4]$ (see Section 3.3.5.2), satisfaction significantly exceeds the non-diversified cases. However, online results also reveal that too much diversification, $\Theta_F \in [0.6, 0.9]$, appears harmful and detrimental to user satisfaction.
- **Accuracy versus satisfaction.** Several researchers have argued that "accuracy does not tell the whole story" [Cosley et al, 2002; Herlocker et al, 2004]. Nevertheless, no evidence has been given to show that some aspects of actual user satisfaction reach beyond accuracy. We close this gap by contrasting our online and offline results, showing that for $\Theta_F \to 0.4$, accuracy deteriorates while satisfaction improves.

4.2   Related Work

Few efforts have addressed the problem of making top-N lists more diverse. Considering literature on collaborative filtering and recommender systems in general only, none have been presented before, to our best knowledge.

However, some work related to our topic diversification approach can be found in information retrieval, specifically meta-search engines. A critical aspect of meta-search engine design is the merging of several top-N lists into one single top-N list. Intuitively, this merged top-N list should reflect the highest quality ranking possible, also known as the "rank aggregation problem" [Dwork et al, 2001]. Most approaches use variations of the "linear combination of score" model (LC), described by Vogt and Cottrell [1999]. The LC model effectively resembles our scheme for merging the original, accuracy-based ranking with the current dissimilarity ranking, but is more general and does not address the diversity issue. Fagin et al [2003] propose metrics for measuring the distance between top-N lists, i.e., inter-list similarity metrics, in order to evaluate the quality of merged ranks. Oztekin et al [2002] extend the linear combination approach by proposing rank combination models that also incorporate content-based features in order to identify the most relevant topics.

More related to our idea of creating lists that represent the whole plethora of the user's topic interests, Kummamuru et al [2004] present their clustering scheme that groups search results into clusters of related topics. The user can then conveniently browse topic folders relevant for his search interest. The commercially available search engine Northern Light (*http://www.northernlight.com*) incorporates similar functionalities. Google (*http://www.google.com*) uses several mechanisms to suppress top-N list items that are too similar in content, showing them only upon the user's explicit request. Unfortunately, no publications on that matter are available.

4.3 Empirical Analysis

We conducted offline evaluations to understand the ramifications of topic diversification on accuracy metrics, and online analysis to investigate how our method affects actual user satisfaction. We applied topic diversification with factors $\Theta_F \in \{0, 0.1, 0.2, \ldots 0.9\}$ to lists generated by both user-based CF and item-based CF, observing effects that occur when steadily increasing Θ_F and analyzing how both approaches respond to diversification.

For online as well as offline evaluations, we used data gathered from BookCrossing (*http://www.bookcrossing.com*). This community caters for book lovers exchanging books around the world and sharing their experiences with other readers.

Data Collection

In a 4-week crawl, we collected data about $278,858$ members of BookCrossing and $1,157,112$ ratings, both implicit and explicit, referring to $271,379$ distinct ISBNs. Invalid ISBNs were excluded from the outset.

The complete BookCrossing dataset, anonymized for data privacy, is available via the author's homepage (*http://www.informatik.uni-freiburg.de/~cziegler/BX/*).

Next, we mined Amazon.com's book taxonomy, comprising 13,525 distinct topics. In order to be able to apply topic diversification, we mined supplementary content information, focusing on taxonomic descriptions that relate books to taxonomy nodes from Amazon.com (*http://www.amazon.com*). Since many books on BookCrossing refer to rare, non-English books, or outdated titles not in print anymore, we were able to garner background knowledge for only 175,721 books. In total, 466,573 topic descriptors were found, giving an average of 2.66 topics per book.

Condensation Steps

Owing to the BookCrossing dataset's extreme sparsity, we decided to *condense* the set in order to obtain more meaningful results from CF algorithms when computing recommendations. Hence, we discarded all books missing taxonomic descriptions, along with all ratings referring to them. Next, we also removed book titles with fewer than 20 overall mentions. Only community members with at least 5 ratings each were kept.

The resulting dataset's dimensions were considerably more moderate, comprising 10,339 users, 6,708 books, and 361,349 book ratings.

4.3.1 Offline Experiments

We performed offline experiments comparing precision, recall, and intra-list similarity scores for 20 different recommendation list setups. Half these lists were based upon user-based CF with different degrees of diversification, the others on item-based CF. Note that we did not compute MAE metric values since we are dealing with implicit rather than explicit ratings.

The before-mentioned *intra-list similarity metric* intends to capture the *diversity* of a list. Diversity may refer to all kinds of features, e.g., genre, author, and other discerning characteristics. Based upon an arbitrary function $c_{ILS} : B \times B \to [-1, +1]$ measuring the similarity $c_{ILS}(b_k, b_e)$ between products b_k, b_e according to some custom-defined criterion, we define intra-list similarity for an agent a_i's list P_{w_i} as follows:

$$\mathrm{ILS}(P_{w_i}) = \frac{\sum_{b_k \in \Im P_{w_i}} \sum_{b_e \in \Im P_{w_i}, b_k \neq b_e} c_{ILS}(b_k, b_e)}{2} \qquad (4.1)$$

Higher metric scores express lower diversity. An interesting mathematical feature of $\mathrm{ILS}(P_{w_i})$ we are referring to in later sections is *permutation-insensitivity*, i.e., let S_N be the symmetric group of all permutations on $N = |P_{w_i}|$ symbols:

$$\forall \sigma_i, \sigma_j \in S_N : \mathrm{ILS}(P_{w_i} \circ \sigma_i) = \mathrm{ILS}(P_{w_i} \circ \sigma_j) \qquad (4.2)$$

Hence, simply rearranging positions of recommendations in a top-N list P_{w_i} does not affect P_{w_i}'s intra-list similarity.

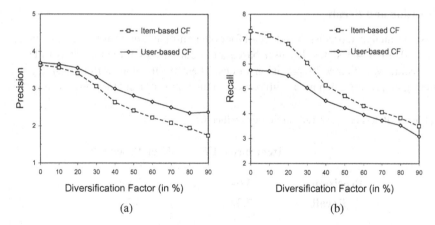

Fig. 4.1 Precision (a) and recall (b) metrics for increasing Θ_F

4.3.1.1 Experiment Setup

For cross-validation of precision and recall metrics of all $10,339$ users, we adopted 4-folding. Hence, rating profiles R_i were effectively split into training sets R_i^x and test sets $T_i^x, x \in \{1, \ldots, 4\}$, at a ratio of $3 : 1$. For each of the $41,356$ different training sets, we computed 20 top-10 recommendation lists.

To generate the diversified lists, we computed top-50 lists based upon pure, i.e., non-diversified, item-based CF and pure user-based CF. The high-performance SUGGEST recommender engine[1] was used to compute these base case lists. Next, we applied the diversification algorithm to both base cases, applying Θ_F factors ranging from 10% up to 90%. For eventual evaluations, all lists were truncated to contain 10 books only.

4.3.1.2 Result Analysis

We were interested in seeing how accuracy, captured by precision and recall, behaves when increasing Θ_F from 0.1 up to 0.9. Since topic diversification may make books with high predicted accuracy trickle down the list, we hypothesized that accuracy will *deteriorate* for $\Theta_F \rightarrow 0.9$. Moreover, in order to find out if our novel algorithm has any significant, positive effects on the diversity of items featured, we also applied our intra-list similarity metric. An overlap analysis for diversified lists, $\Theta_F \geq 0.1$, versus their respective non-diversified pendants indicates how many items stayed the same for increasing diversification factors.

---

[1] Visit `http://www-users.cs.umn.edu/~karypis/suggest/` for further details.

Precision and Recall

First, we analyzed precision and recall for both non-diversified base cases, i.e., when $\Theta_F = 0$. Table 4.1 states that user-based and item-based CF exhibit almost identical accuracy, indicated by precision values. Their recall values differ considerably, hinting at deviating behavior with respect to the types of users they are scoring for.

Table 4.1 Precision and recall for non-diversified CF

	Item-based CF	User-based CF
Precision	3.64	3.69
Recall	7.32	5.76

Next, we analyzed the behavior of user-based and item-based CF when steadily increasing Θ_F by increments of 10%, depicted by Figure 4.1. The two charts reveal that diversification has detrimental effects on *both* metrics and on *both* CF algorithms. Interestingly, corresponding precision and recall curves have almost identical shape.

The loss in accuracy is more pronounced for item-based than for user-based CF. Furthermore, for either metric and either CF algorithm, the drop is most distinctive for $\Theta_F \in [0.2, 0.4]$. For lower Θ_F, negative impacts on accuracy are marginal. We believe this last observation due to the fact that precision and recall are permutation-insensitive, i.e., the mere order of recommendations within a top-N list does not influence the metric value, as opposed to Breese score (see Section 2.4.1.2). However, for low Θ_F, the pressure that the dissimilarity rank exerts on the top-N list's makeup is still too weak to make many new items diffuse into the top-N list. Hence, we conjecture that rather the *positions* of current top-N items change, which does not affect either precision or recall.

Intra-list Similarity

Knowing that our diversification technique bears a significant, *negative* impact on accuracy metrics, we wanted to know how our approach affected the intra-list similarity measure. Similar to the precision and recall experiments, we computed metric values for user-based and item-based CF with $\Theta_F \in [0, 0.9]$ each. We instantiated the metric's embedded similarity function c_{ILS} with our taxonomy-driven metric c^*, defined in Section 3.3.5.2. Results obtained are provided by Figure 4.2.

The topic diversification method considerably lowers the pairwise similarity between list items, thus making top-N recommendation lists more diverse. Diversification appears to affect item-based CF stronger than its user-based counterpart, in line with our findings about precision and recall. For lower Θ_F, curves are less steep than for $\Theta_F \in [0.2, 0.4]$, which also well aligns with our precision and recall analysis. Again, the latter phenomenon can be explained by one of the metric's inherent features: like precision and recall, intra-list similarity is permutation-insensitive.

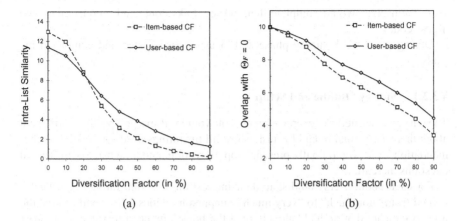

Fig. 4.2 Intra-list similarity (a) and original list overlap (b) for increasing Θ_F

Original List Overlap

The right-hand side of Figure 4.2 depicts the number of recommended items staying the same when increasing Θ_F with respect to the original list's content. Both curves exhibit roughly linear shapes, being less steep for low Θ_F, though. Interestingly, for factors $\Theta_F \leq 0.4$, at most 3 recommendations change on average.

Conclusion

We found that diversification appears largely detrimental to both user-based and item-based CF along precision and recall metrics. In fact, this outcome aligns with our expectations, considering the nature of those two accuracy metrics and the way that the topic diversification method works. Moreover, we found that item-based CF seems more susceptible to topic diversification than user-based CF, backed by results from precision, recall and intra-list similarity metric analysis.

4.3.2 User Survey

Offline experiments helped us in understanding the implications of topic diversification on both CF algorithms. We could also observe that the effects of our approach are different on different algorithms. However, knowing about the deficiencies of accuracy metrics, we wanted to assess *real* user satisfaction for various degrees of diversification, thus necessitating an online survey.

For the online study, we computed each recommendation list type anew for users in the denser BookCrossing dataset, though without K-folding. In cooperation with BookCrossing, we mailed all eligible users via the community mailing system, asking them to participate in our online study. Each mail contained a personal link that would direct the user to our online survey pages. In order to make sure that only the users themselves would complete their survey, links contained unique, encrypted access codes.

During the 3-week survey phase, 2,125 users participated and completed the study.

4.3.2.1 Survey Outline and Setup

The survey consisted of several screens that would tell the prospective participant about this study's nature and his task, show all his ratings used for making recommendations, and would finally present a top-10 recommendation list, asking several questions thereafter.

For each book, users could state their interest on a 5-point rating scale. Scales ranged from "not much" to "very much", mapped to values 1 to 4, and offered the user to indicate that he had "already read the book", mapped to value 5. In order to successfully complete the study, users were *not* required to rate all their top-10 recommendations. Neutral values were assumed for non-votes instead. However, we required users to answer all further questions, concerning the list as a whole rather than its single recommendations, before submitting their results. We embedded those questions we were actually keen about knowing into ones of lesser importance, in order to conceal our intentions and not bias users.

The one top-10 recommendation list for each user was chosen among 12 candidate lists, either user-based or item-based featuring no diversification, i.e., $\Theta_F = 0$, medium levels, $\Theta_F \in \{0.3, 0.4, 0.5\}$, and high diversification, $\Theta_F \in \{0.7, 0.9\}$. We opted for those 12 instead of all 20 list types in order to acquire enough users completing the survey for each slot. The assignment of a specific list to the current user was done dynamically, at the time of the participant entering the survey, and in a round-robin fashion. Thus, we could guarantee that the number of users per list type was roughly identical.

4.3.2.2 Result Analysis

For the analysis of our inter-subject survey, we were mostly interested in the following three aspects. First, the *average rating* users gave to their 10 single recommendations. We expected results to roughly align with scores obtained from precision and recall, owing to the very nature of these metrics. Second, we wanted to know if users perceived their list as well-diversified, asking them to tell whether the lists reflected rather a broad or narrow *range of their reading interests*. Referring to the intra-list similarity metric, we expected the users' perceived range of topics, i.e., the list's diversity, to increase with increasing Θ_F. Third, we were curious about the *overall satisfaction* of users with their recommendation lists in their entirety, the measure to compare performance.

Both last-mentioned questions were answered by each user on a 5-point likert scale, higher scores denoting better performance. Moreover, we averaged the

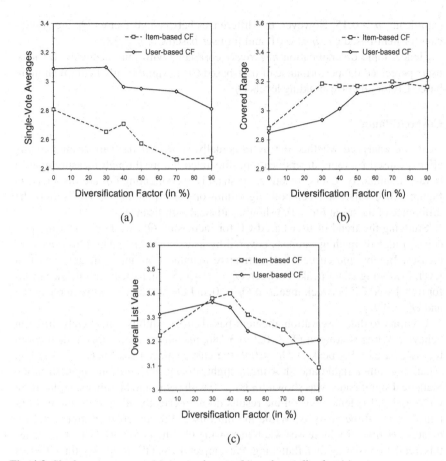

Fig. 4.3 Single-vote averages (a), covered range (b), and overall value (c)

eventual results by the number of users. Statistical significance of all mean values was measured by parametric one-factor ANOVA (see, e.g., [Armitage and Berry, 2001]), where $p < 0.05$ is assumed if not indicated otherwise.

Single-Vote Averages

Users perceived recommendations made by user-based CF systems on average as more accurate than those made by item-based CF systems, as depicted in Figure 4.3(a). At each featured diversification level Θ_F, the differences between the two CF types are statistically significant, $p \ll 0.01$.

Moreover, for each algorithm, higher diversification factors obviously entail lower single-vote average scores, which confirms our hypothesis stated before. The item-based CF's cusp at $\Theta_F \in [0.3, 0.5]$ appears as a notable outlier, opposed to the trend, but differences between the 3 means at $\Theta_F \in [0.3, 0.5]$ are not statistically significant, $p > 0.15$. However, the differences between all factors Θ_F are significant for item-based CF, $p \ll 0.01$, and for user-based CF, $p < 0.1$.

Hence, topic diversification *negatively* correlates with pure accuracy. Besides, users perceived the performance of user-based CF as significantly better than item-based CF for all corresponding levels Θ_F.

Covered Range

Next, we analyzed whether the users actually *perceived* the variety-augmenting effects caused by topic diversification, illustrated before through measurement of intra-list similarity. Users' reactions to steadily incrementing Θ_F are displayed in Figure 4.3(b). First, between both algorithms on corresponding Θ_F levels, only the difference of means at $\Theta_F = 0.3$ shows statistical significance.

Studying the trend of user-based CF for increasing Θ_F, we notice that the perceived range of reading interests covered by users' recommendation lists also increases. Hereby, the curve's first derivative maintains an approximately constant level, exhibiting slight peaks between $\Theta_F \in [0.4, 0.5]$. Statistical significance holds for user-based CF between means at $\Theta_F = 0$ and $\Theta_F > 0.5$, and between $\Theta_F = 0.3$ and $\Theta_F = 0.9$.

Contrary to that observation, the item-based curve exhibits a drastically different behavior. While soaring at $\Theta_F = 0.3$ to 3.186, reaching a score almost identical to the user-based CF's peak at $\Theta_F = 0.9$, the curve barely rises for $\Theta_F \in [0.4, 0.9]$, remaining rather stable and showing a slight, though insignificant, upward trend. Statistical significance was shown for $\Theta_F = 0$ with respect to all other samples from $\Theta_F \in [0.3, 0.9]$. Hence, our online results do not perfectly align with findings obtained from offline analysis. While the intra-list similarity chart in Figure 4.2 indicates that diversity increases when increasing Θ_F, the item-based CF chart defies this trend, first soaring then flattening. We conjecture that the following three factors account for these peculiarities:

Diversification factor impact. Offline results for the intra-list similarity metric already indicated that the impact of topic diversification on item-based CF is much stronger than on user-based CF. Consequently, the item-based CF's user-perceived interest coverage is significantly higher at $\Theta_F = 0.3$ than the user-based CF's.

Human perception. We believe that human perception can capture the level of diversity inherent to a list only to some extent. Beyond that point, increasing diversity remains unnoticed. For the application scenario at hand, Figure 4.3 suggests this point around score value 3.2, reached by user-based CF only at $\Theta_F = 0.9$, and approximated by item-based CF already at $\Theta_F = 0.3$.

Interaction with accuracy. Analyzing results obtained, we have to bear in mind that covered range scores are *not* fully independent from single-vote averages. When accuracy is poor, i.e., the user feels unable to identify recommendations that are interesting to him, chances are high his discontentment will also negatively affect his diversity rating. For $\Theta_F \in [0.5, 0.9]$, single-vote averages are remarkably low, which might explain why perceived coverage scores do not improve for increasing Θ_F.

However, we may conclude that users *do* perceive the application of topic diversification as an overly positive effect on reading interest coverage.

Overall List Value

The third feature variable we were evaluating, the overall value users assigned to their personal recommendation list, effectively represents the "target value" of our studies, measuring actual user satisfaction. Owing to our conjecture that user satisfaction is a mere composite of accuracy and other influential factors, such as the list's diversity, we hypothesized that the application of topic diversification would *increase* satisfaction. At the same time, considering the downward trend of precision and recall for increasing Θ_F, in accordance with declining single-vote averages, we expected user satisfaction to drop off for large Θ_F. Hence, we supposed an arc-shaped curve for both algorithms.

Results for the overall list value are provided by Figure 4.3(c). Analyzing user-based CF, we observe that the curve does *not* follow our hypothesis. Slightly improving at $\Theta_F = 0.3$ over the non-diversified case, scores drop for $\Theta_F \in [0.4, 0.7]$, eventually culminating in a slight but visible upturn at $\Theta_F = 0.9$. While lacking reasonable explanations and being opposed to our hypothesis, the curve's data-points actually bear no statistical significance for $p < 0.1$. Hence, we conclude that topic diversification has a marginal, largely negligible impact on overall user satisfaction, initial positive effects eventually being offset by declining accuracy.

On the contrary, for item-based CF, results obtained look very different. In compliance with our previous hypothesis, the curve's shape roughly follows an arc, peaking at $\Theta_F = 0.4$. Taking the three data-points defining the arc, we obtain statistical significance for $p < 0.1$. The endpoint's score at $\Theta_F = 0.9$ being inferior to the non-diversified case's, we observe that too much diversification appears detrimental, most likely owing to substantial interactions with accuracy.

Eventually, for overall list value analysis, we come to conclude that topic diversification has no measurable effects on user-based CF, but significantly improves item-based CF performance for diversification factors Θ_F around 40%.

4.3.2.3 Multiple Linear Regression

Results obtained from analyzing user feedback along various feature axes already indicated that users' overall satisfaction with recommendation lists not only depends on accuracy, but also on the range of reading interests covered. In order to more rigidly assess that indication by means of statistical methods, we applied *multiple linear regression* to our survey results, choosing the overall list value as dependent variable. As independent input variables, we provided single-vote averages and covered range, both appearing as first-order and second-order polynomials, i.e., SVA and CR, and SVA^2 and CR^2, respectively. We also tried several other, more complex models, without achieving significantly better model fitting.

Analyzing multiple linear regression results, shown in Table 4.2, confidence values $P(>|t|)$ clearly indicate that statistically significant correlations for accuracy and covered range with user satisfaction exist.

Since statistical significance also holds for their respective second-order polynomials, i.e., CR^2 and SVA^2, we conclude that these relationships are non-linear and more complex, though.

As a matter of fact, linear regression delivers a strong indication that the intrinsic utility of a list of recommended items is more than just the average value of accuracy votes for all single items, but also depends on the perceived diversity.

Table 4.2 Multiple linear regression results

	Estimate	Error	t-Value	$P(>|t|)$
(const)	3.27	0.023	139.56	$< 2e - 16$
SVA	12.42	0.973	12.78	$< 2e - 16$
SVA^2	-6.11	0.976	-6.26	$4.76e - 10$
CR	19.19	0.982	19.54	$< 2e - 16$
CR^2	-3.27	0.966	-3.39	0.000727

Multiple R^2: 0.305, adjusted R^2: 0.303

4.3.3 Limitations

There are some limitations to the study, notably referring to the way topic diversification was implemented. Though the Amazon.com taxonomies were human-created, there might still be some mismatch between what the topic diversification algorithm perceives as "diversified" and what humans do. The issue is effectively inherent to the taxonomy's structure, which has been designed with *browsing tasks* and ease of searching rather than with interest profile generation in mind. For instance, the taxonomy features topic nodes labelled with letters for alphabetical ordering of authors from the same genre, e.g., BOOKS → FICTION → ... → AUTHORS, A-Z → G. Hence, two Sci-Fi books from two different authors with the same initial of their last name would be classified under the same node, while another Sci-Fi book from an author with a *different* last-name initial would *not*. Though the problem's impact is largely marginal, owing to the relatively deep level of nesting where such branchings occur, the procedure appears far from intuitive.

An alternative approach to further investigate the accuracy of taxonomy-driven similarity measurement, and its limitations, would be to have *humans* do the clustering, e.g., by doing card sorts or by estimating the similarity of any two books contained in the book database. The results could then be matched against the topic diversification method's output.

4.4 Summary

This chapter provided empirical analyses in order to evaluate the application of our topic diversification method to common collaborative filtering algorithms, and introduced the novel *intra-list similarity* metric.

Contrasting precision and recall metrics, computed both for user-based and item-based CF and featuring different levels of diversification, with results obtained from a large-scale user survey, we showed that the user's overall liking of recommendation lists goes beyond accuracy and involves other factors, e.g., the users' perceived list diversity. We were thus able to provide empirical evidence that lists are *more* than mere aggregations of single recommendations, but bear an intrinsic, added value.

Though effects of diversification were largely marginal on user-based CF, item-based CF performance improved significantly, an indication that there are some behavioral differences between both CF classes. Moreover, while pure item-based CF appeared slightly inferior to pure user-based CF in overall satisfaction, diversifying item-based CF with factors $\Theta_F \in [0.3, 0.4]$ made item-based CF outperform user-based CF. Interestingly, for $\Theta_F \leq 0.4$, no more than three items changed with respect to the original list, shown in Figure 4.2. Small changes thus have high impact.

We believe our findings especially valuable for practical application scenarios, knowing that many commercial recommender systems, eg., Amazon.com [Linden et al, 2003] and TiVo [Ali and van Stam, 2004], are item-based, owing to the algorithm's computational efficiency. For these commercial systems, topic diversification could be an interesting supplement, increasing user satisfaction and thus the customers' incentive to purchase recommended goods.

Chapter 5
Taxonomies for Calculating Semantic Proximity

"For just when ideas fail, a word comes in to save the situation."

– Johann Wolfgang von Goethe (1749 – 1832)

5.1 Introduction

The two preceding chapters have demonstrated that classification taxonomies can be put to use in improving recommender systems in terms of the *quality* of their recommendations. The taxonomy we resorted to for all the empirical evaluations was the one from Amazon.com. Now we want to give an example how *Web 2.0* taxonomies, having been crafted by collective efforts of several thousands of volunteering editors, can likewise be used to these ends.

The research we present in this chapter deals with an approach of how these Web 2.0 taxonomies can be exploited to assess the semantic proximity of two concepts, represented by n-grams each. While this scenario does not directly address this book's dominating topic of improving recommender systems, there are clearly links to it: Content-based recommenders rely on matching descriptive features of products against each other, in order to determine the similarity of these products for the calculation of the item-item similarity matrix. Now, these product features often include textual information. Being able to compute the semantic proximity of two syntactically different n-grams (one occurring in the description of the one product, one occurring in the other's) will allow to improve the quality of this product-product similarity metric, and thus the quality of the recommender itself.

5.1.1 On Calculating the Similarity of Word Meanings

Research on similarity of word meanings dates back to the early 60's, see Rubenstein and Goodenough [1965]. Thenceforward, numerous semantic proximity measures have been proposed, mostly operating on the taxonomic dictionary WordNet [Miller, 1995; Resnik, 1995; Lin, 1998; Budanitsky and Hirst, 2000; Li et al, 2003],

C.-N. Ziegler: *Social Web Artifacts for Boosting Recommenders*, SCI 487, pp. 61–77.
DOI: 10.1007/978-3-319-00527-0_5 © Springer International Publishing Switzerland 2013

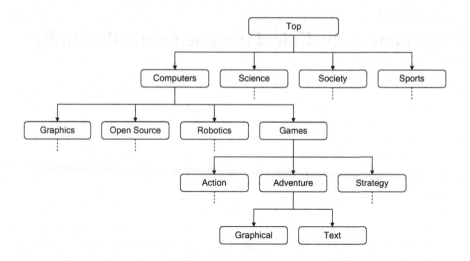

Fig. 5.1 Extracted fragment from the ODP taxonomy

exploiting its hierarchical structuring. The main objective of these approaches is to mimic human judgement with respect to the relatedness of two concepts, e.g., BI-CYCLES and CARS. With the advent of applications that intend to make machines understand and extract *meaning* from human-crafted information, e.g., the Semantic Web initiative or text mining, the necessity for tools enabling the automatic detection of semantic proximity becomes even stronger. However, one severe drawback of these approaches is that their application has been confined to WordNet only. While the number of unique strings of this taxonomically organized dictionary, i.e., nouns, verbs, adjectives, and adverbs, nears 150,000, large amounts of information available on the Web and other textual sources cannot be captured by such dictionaries. Particularly brand names, names of artists, locations, products and composed terms, in other words, specific *instances* of concepts, are beyond their scope. Examples for these named entities are CIKM, DATABASE THEORY, NEIL ARMSTRONG, or XBOX GAMES, to name some.

5.1.2   Contributions

We intend to overcome the aforementioned issue and propose an architecture that allows to compute the estimated semantic proximity between *arbitrary* concepts and concept instances.[1] The following two major contributions are made:

- **Framework leveraging Google and ODP.** Instead of merely exploiting Word-Net, we leverage the power of the entire Web and the endeavors of thousands

---

[1] We will abuse language by likewise denoting *concepts*, e.g., POET, and *instances*, e.g., FRIEDRICH SCHILLER, by the term *concept* only.

of voluntary human editors who classify Web pages into the ODP taxonomy, also known as DMOZ. Google serves as an instrument to obtain Web pages that match one particular concept, e.g., FRIEDRICH SCHILLER. The Open Directory Project (ODP) then allows us to classify these result pages into a human-crafted taxonomy. Thus, we are able to garner a *semantic profile* of each concept.

- **Metric for multi-class categorization.** Most taxonomy-based proximity measures are geared towards computing the semantic distance of word senses that fall into exactly *one* category each. We propose a metric that allows a multi-class approach (see Ziegler et al [2004a] and Chapter 3) where each given concept may be arranged into several categories, which is essential for our architectural setup. Empirical evaluation through an online user study demonstrates the superior performance of our approach against traditional similarity metrics.

5.2 Related Work

The study of semantic proximity between two given concepts has largely focused on *similarity*, e.g., synonymy and hyponymy [Miller, 1995]. Proximity goes even further, also subsuming meronymy (part-whole) and arbitrarily typed semantic relationships.

Early taxonomy-based similarity metrics only have taken into account the shortest path ϕ between two concepts within the taxonomy, and the depth σ of the most specific common subsumer of both concepts. See Budanitsky and Hirst [2000] for an overview of these early works. Next-generation approaches were inspired by information theory and only used taxonomies in combination with text corpora. Thus, the probability of a concept, its so-called *information content*, could be computed and used to refine the similarity measure. Resnik [1995, 1999] lays the foundations of this approach, followed by Jiang and Conrath [1997] and Lin [1998], both largely similar to Resnik's. Lin's approach is extended to handle graphs rather than mere trees, as proposed by Maguitman et al [2005].

Li et al [2003] have conducted an extensive study that revealed that the usage of information content does not yield better performance. Moreover, they proposed a metric that combines shortest path length ϕ and subsumer depth σ in a *non-linear* fashion and outperformed traditional taxonomy-based approaches. Chirita et al [2005] used a variation of Li et al's metric for personalizing Web search.

Taxonomy-based metrics with collaborative filtering systems in mind have been proposed by Ganesan et al [2003] and Ziegler et al [2004a].

The exploitation of Web search engine results for computing similarity between concepts or, more generally, queries, has been tried before. Chien and Immorlica [2005] and Vlachos et al [2004] attempted to detect similarity via *temporal* correlation, reporting mixed results. Wen et al [2001] computed semantic query similarity based on query content, user feedback, and some simple document hierarchy. No empirical analyses were provided, though. Cimiano et al [2004, 2005] make use of linguistic patterns along with search engine leverage to automatically identify class-instance relationships and disambiguate word senses.

5.3   Framework

In this section, we describe the framework we built in order to compute semantic
proximity between arbitrary concepts or instances. The approach rests upon ODP
and Google Directory, which we use in order to provide us with the required background knowledge. The two services are required so we can compose *semantic profiles* of the concepts we want to compare. The following step then necessitates a
proximity metric to match these profiles against each other.

5.3.1   Service Requirements

Our architecture is based on mainly two pillars:

- **Open Directory Project.** The Open Directory Project (*http://www.dmoz.org*),
  also called DMOZ, combines the joint efforts of more than 69,000 volunteering
  editors helping to categorize the Web. ODP is the largest and most comprehen-
  sive human-edited Web page catalog currently available. Organized as a tree-
  structured taxonomy, 590,000 categories build the inner nodes of the directory.
  Leaf nodes are given by more than 5.1 million sites that have been categorized
  into the ODP already.
- **Google Directory.** Google Directory (*http://www.google.com/dirhp*) runs on top
  of ODP and provides search results with additional *referrals* into the ODP, thus
  extending the traditional Google service. These referrals represent paths from
  the ODP's inner nodes or leafs to its root node, $\top$. For one given query, only
  those pages are returned as results that have been categorized into the ODP. For
  example, the ODP referral that Google Directory's first search result assigns to
  the ACM main page (*http://www.acm.org*) looks as follows:

$$\top \to \text{COMPUTERS} \to \text{COMPUTER SCIENCE} \to \text{ORGANIZATIONS} \ldots \qquad (5.1)$$

5.3.2   Algorithm Outline

The main task is to compute the proximity s between two concepts c_x, c_y, i.e.,
$s(c_x, c_y)$. In order to match c_x against c_y, we need to build semantic profiles for
both concepts first.

To this end, we send these two concepts c_x and c_y, for instance BORIS BECKER
and WIMBLEDON, to Google Directory and obtain two *ranked result lists* $q^{c_x} : \mathscr{L}^{n_x}$
and $q^{c_y} : \mathscr{L}^{n_y}$, respectively. We define $\mathscr{L}^{n_z} := \{1, 2, \ldots, n_z\} \to D$, where $z \in \{x, y\}$,
n_z the number of documents returned for query term c_z, and D the set of topics
in the ODP taxonomy. Hence, we only consider the *ODP referral* associated with
each document returned for the query rather than the textual summary. For example,
$q^{c_x}(1)$ gives the ODP topic that the top-ranked result document for query term c_x is
categorized into.

Next, the two profiles are forwarded to the proximity metric, which then computes the estimated semantic proximity score $s(c_x, c_y)$, using the ODP as background knowledge to look up topics $q^{c_z}(i)$, where $z \in \{x, y\}$ and $i \in \{1, 2, \ldots, n_z\}$, within the taxonomy.

The approach is very versatile and not only extends to the computation of semantic proximity for pairs of concepts, but effectively pairs of arbitrary queries in general.

5.4 Proximity Metrics

In order to use our framework, we need to install an actual taxonomy-based proximity metric s that compares q^{c_x} with q^{c_y}. Since q^{c_x} and q^{c_y} are ranked lists of topics, its type must be $s : (\mathscr{L}^{n_x} \times \mathscr{L}^{n_y}) \to S$, where S is an arbitrary scale, e.g., $[-1, +1]$.

We will first review several popular metrics that have been proposed in the context of WordNet and then proceed to propose our own taxonomy-based proximity metric.

5.4.1 Similarity between Word-Sense Pairs

In general, WordNet metrics compare the similarity of *word senses*, where each word sense is represented by a topic from the taxonomy. Hence, the metric's functional layout is $s : D \times D \to S$ rather than $s : (\mathscr{L}^{n_x} \times \mathscr{L}^{n_y}) \to S$. The reason is that we are comparing two *topics* from the taxonomy rather than *instances*, which are arranged into multiple topics within the taxonomy. For example, BATMAN is an instance of *several* topics, e.g. MOVIE and CARTOON HERO. We will show later on how to circumvent this issue.

The simplest WordNet metric only computes the shortest path $\phi(d_x, d_y)$ between two topics d_x, d_y in the taxonomy. Leacock and Chodorow (see [Budanitsky and Hirst, 2000]) modify this basic metric by scaling the path length by the overall depth λ of the taxonomy:

$$s_{LC}(d_x, d_y) = -\log\left(\frac{\phi(c_x, c_y)}{2 \cdot \lambda}\right) \tag{5.2}$$

Though the Leacock-Chodorow metric uses little information to compute the similarity estimate, its accuracy has been shown only insignificantly inferior to more informed approaches based on information theory, e.g., Resnik [1995, 1999] or Jiang and Conrath [1997].

Li et al [2003] have conducted an extensive survey comparing numerous existing WordNet metrics and proposed a new metric which combines shortest path length $\phi(d_x, d_y)$ and most specific subsumer depth $\sigma(d_x, d_y)$ in a *non-linear* fashion, outperforming all other benchmark metrics. The most specific subsumer of topics d_x and d_y is defined as the topic that lies on the taxonomy paths from the root topic to both d_x and d_y and has maximal distance from the root topic. Li et al's metric also features two tuning parameters α and β which have to be learned in order to guarantee the metric's optimal performance. The metrics is defined as follows:

$$s_{LBM}(d_x, d_y) = e^{-\alpha \cdot \phi(d_x, d_y)} \cdot \frac{e^{\beta \cdot \sigma(d_x, d_y)} - e^{-\beta \cdot \sigma(d_x, d_y)}}{e^{\beta \cdot \sigma(d_x, d_y)} + e^{-\beta \cdot \sigma(d_x, d_y)}} \qquad (5.3)$$

As has been stated before, the above metrics only measure the distance between two singleton topics d_x and d_y rather than two lists of topics $q^{c_x} : \mathscr{L}^{n_x}$ and $q^{c_y} : \mathscr{L}^{n_y}$, respectively. The issue has been addressed by Chirita et al [2005] by computing the average similarity of all unordered pairs $\{d_x, d_y\} \in \mathfrak{I}(q^{c_x}) \times \mathfrak{I}(q^{c_y})$, where $d_x \neq d_y$.[2]

5.4.2 Multi-class Categorization Approach

As outlined in Section 5.3.2, multi-class categorization of concepts/instances into *several* topics is essential for our approach, since more than one query result and its taxonomic referral are used to describe one concept/instance c_z. Existing WordNet metrics can be tailored to support proximity computations for topic lists, but their performance is non-optimal (see Section 5.5). We therefore propose a new metric with multi-class categorization in mind. Our approach is substantially different from existing metrics and can be subdivided into two major phases, namely *profiling* and *proximity computation*. The first phase takes the list of topics describing one concept/instance c_z, e.g., BATMAN BEGINS, and creates a flat profile vector, based on the ODP taxonomy as background knowledge. The second phase then takes the profile vectors for both c_x and c_y and matches them against each other, hence computing their correlation.

As input, our metric expects two ranked topics lists $q^{c_z} : \mathscr{L}^{n_z}, z \in \{x, y\}$, and three fine-tuning parameters, α, γ, and δ. These parameters have to be learned from a training set before applying the metric.

5.4.2.1 Profiling Phase

For each concept c_z for which to build its profile, we create a new vector $\mathbf{v}_z \in \mathbb{R}^{|D|}$, i.e., the vector's dimension is exactly the number of topics in the ODP taxonomy. Next, we accord a certain score $\tilde{\mu}_i$, where $i \in \{1, 2, \ldots, n_z\}$, for each topic $q^{c_z}(i) \in \mathfrak{I}(q_z)$. The amount of score depends on the *rank* i of topic $q^{c_z}(i)$. The earlier the topic appears in the result list q^{c_z} of query c_z, the more weight we assign to that topic, based upon the assumption that results further down the list are not as valuable as top-list entries. For the weight assignment, we assume an *exponential decay*, inspired by Breese et al's half-life utility metric (see Breese et al [1998]):

$$\tilde{\mu}_i = 2^{-(i-1)/(\alpha-1)} \qquad (5.4)$$

Parameter α denotes the *impact weight half-life*, i.e., the number of the rank of topic $q^{c_z}(\alpha)$ on list q^{c_z} for which the impact weight is exactly half as much as the weight $\tilde{\mu}_1$ of the top-ranked result topic. When assuming $\alpha = \infty$, all ranks are given equal weight.

---

[2] $\mathfrak{I}(f)$ denotes the *image* of map $f : A \to B$, i.e., $\mathfrak{I}(f : A \to B) := \{f(x) \mid x \in A\}$.

Having computed the score $\tilde{\mu}_i$ for each topic $q^{c_z}(i)$, we now start to assign score to all topics $d_{i,0}, d_{i,1}, \ldots, d_{i,\lambda(i)}$ along the path from $q^{c_z}(i)$ to the taxonomy's root node. Hereby, $\lambda(i)$ denotes the *depth* of topic $q^{c_z}(i) = d_{i,\lambda(i)}$ in our taxonomy, and $d_{i,0}$ is the root node. The idea is to propagate score from leaf topic $d_{i,\lambda(i)}$ to all its ancestors, for each $d_{i,j}$ is also "a type of" $d_{i,j-1}$, owing to the taxonomy's nature of being composed of hierarchical "is-a" relationships.

Note that the ODP taxonomy also features some few links of types other than "is-a", namely "symbolic" and "related". These types were not considered in our model so far. When upward-propagating score for each $q^{c_z}(i)$, we first assign score $\mu_{i,\lambda(i)} := \tilde{\mu}_i$ to $d_{i,\lambda(i)}$. The score for its parent topic $d_{i,\lambda(i)-1}$ then depends on four factors, namely parameters γ and δ, the number of siblings of $d_{i,\lambda(i)}$, denoted $\rho(d_{i,\lambda(i)})$, and the score $\mu_{i,\lambda(i)}$ of $d_{i,\lambda(i)}$. The general score propagation function from taxonomy level j to level $j-1$ is given as follows:

$$\mu_{i,j-1} = \mu_{i,j} \cdot \frac{1}{\gamma + \delta \cdot \log(\rho(d_{i,j}) + 1)} \tag{5.5}$$

Informally, the propagated score depends on a constant factor γ and the number of siblings that topic $d_{i,j}$ has. The more siblings, the less score is propagated upwards. In order to not overly penalize nodes $d_{i,j-1}$ that have numerous children, we chose *logarithmic* scaling. Parameter δ controls the influence that the number of siblings has on upward propagation. Clearly, other functions could be used likewise.

Next, we normalize all score $\mu_{i,j}$, where $i \in \{1, 2, \ldots, n_z\}$ and $j \in \{0, 1, \ldots, \lambda(i)\}$, so that values $\mu_{i,j}$ sum up to unit score. Values of vector $\mathbf{v}_z$ at positions $d_{i,j}$ are eventually increased by $\mu_{i,j}$, yielding the final *profile vector* for concept c_z.

The algorithm shown in Alg. 5.1 summarizes the complete profiling procedure. Function pathvec$(q^{c_z}(i) \in D)$ returns the vector containing the path of topic $q^{c_z}(i)$'s ancestors, from $q^{c_z}(i)$ itself to the root node. The vector's size is $\lambda(i)$. Function id$(d \in D)$ gives the *index* that topic d is mapped to within profile vector $\mathbf{v}_z$.

5.4.2.2 Measuring Proximity

Profile generation for concepts c_x, c_y and their respective ranked topic lists q^{c_x}, q^{c_y} appears as the major task of our approach; the eventual proximity computation is straight-forward. Mind that the profiling procedure generates *plain feature vectors*, so we can apply generic statistical tools for measuring vector similarity. We opted for Pearson's correlation coefficient, particularly prominent in collaborative filtering applications [Sarwar et al, 2001; Ziegler et al, 2005]. Hence, the final proximity value is computed as follows:

$$s_{ZSL}(\mathbf{v}_x, \mathbf{v}_y) = \frac{\sum\limits_{k=0}^{|D|} (v_{x,k} - \bar{v}_x) \cdot (v_{y,k} - \bar{v}_y)}{\left(\sum\limits_{k=0}^{|D|} (v_{x,k} - \bar{v}_x)^2 \cdot \sum\limits_{k=0}^{|D|} (v_{y,k} - \bar{v}_y)^2 \right)^{\frac{1}{2}}} \tag{5.6}$$

Where $\bar{v}_x$ and $\bar{v}_y$ denote the mean values of vectors $\mathbf{v}_x$ and $\mathbf{v}_y$. Moreover, $\mathbf{v}_x$ and $\mathbf{v}_y$ are assumed to have been computed according to the algorithm in Alg. 5.1.

```
func prof (q^{c_z} : \mathscr{L}^{n_z}, \alpha, \gamma, \delta) returns v_z \in \mathbb{R}^{|D|} {

    set v_z \leftarrow 0, n \leftarrow 0;

    for i \leftarrow 1 to n_z do

        set \tilde{\mu}_i \leftarrow 2^{-(i-1)/(\alpha-1)};

        for j \leftarrow \lambda(i) to 0 do

            set d_{i,j} \leftarrow (pathvec(q^{c_z}(i)))_j;

            if ( j = \lambda(i) ) then
                set \mu_{i,j} \leftarrow \tilde{\mu}_i;
            else
                set \mu_{i,j} \leftarrow \mu_{i,j+1} \cdot (\gamma + \delta \cdot \log(\rho(d_{i,j}+1)))^{-1};
            end if

            set v_{z,id(d_{i,j})} \leftarrow v_{z,id(d_{i,j})} + \mu_{i,j};
            set n \leftarrow n + \mu_{i,j};

        end do
    end do

    for i \leftarrow 1 to |D| do
        set v_{z,i} \leftarrow v_{z,i} / n;
    end do

    return v_z;
}
```

Alg. 5.1 Profiling algorithm

5.5 Empirical Evaluation

In Section 5.3, we have proposed a framework to compute semantic proximity between *arbitrary* concepts/instances. In order to evaluate which proximity metric best fits our approach, we conducted an extensive empirical study involving 51 human subjects and necessitating the creation of two novel benchmark sets, featuring 30 and 25 concept pairs.

The evaluation method follows the methodology used for comparing the performance of WordNet metrics, e.g., Resnik [1995], Lin [1998], Budanitsky and Hirst [2000], and Li et al [2003], based on mainly two benchmark sets, namely Rubenstein and Goodenough [1965] and Miller and Charles [1991]. The first set features 65 concept pairs, e.g., ROOSTER VS. VOYAGE, FURNACE VS. STOVE, and so forth. Miller-Charles is a mere subset of Rubenstein-Goodenough and only contains 30 word pairs. These 30 word pairs were given to a group of 38 people, asking them to judge the semantic similarity of each pair of words on a 5-point scale [Miller and Charles, 1991]. For benchmarking, these human ratings were used as an "arbiter" for all WordNet metrics, and the correlation for each metric's computed word/concept pair similarities with human judgement was measured. The closer the metric's results, the better its accuracy.

5.5.1 Benchmark Layout

Neither Miller-Charles' nor Goodenough-Rubenstein's pair set features concept instances or composed concepts, e.g., locations, book titles, names of actors, etc.; however, the comparison of semantic proximity for these specific terms represents the core capability of our framework. We hence needed to create own benchmark sets. Since some of the metrics presented in Section 5.4, namely Li et al's metric (see [Li et al, 2003]) and our own approach, require parameter learning, we created *two* lists of concept pairs. The first, denoted B_0, contains 25 pairs of concepts, e.g., FOOD NETWORK vs. FLOWERS, or IMDB vs. BLOCKBUSTER, and serves for training and parameter learning. The second benchmark, denoted B_1, features 30 concept pairs such as EASYJET vs. CHEAP FLIGHTS and HOLIDAY INN vs. VALENTINE'S DAY. The conception of both benchmark sets and their respective human subject studies is identical. They only vary in their concept pair lists, which are disjoint from each other. The sets, along with the average ratings of human subjects per concept pair and the respective standard deviations, are given in Table 5.1 and 5.2.

In order to obtain these two lists of concept pairs, we used Google's Suggest service (*http://www.google.com/webhp?complete=1*), still in its beta version at the time of this writing: For each letter in the alphabet (A-Z), we collected the list of most popular search queries proposed by Google Suggest, giving us 260 different queries, e.g., CHEAP FLIGHTS, INLAND REVENUE, and CARS. We took all 33,670 possible query pair combinations and computed the semantic proximity for each query according to our proposed framework, using the simple Leacock-Chodorow metric (see [Budanitsky and Hirst, 2000]). Next, for the test set B_1, we sorted all pairs according to their metric score in descending order and randomly picked ten pairs from the 5% most similar concept pairs, ten concept pairs from mid-range, and ten pairs from the bottom. For B_0, we proceeded in a similar fashion.

We still had to manually weed out unsuitable pairs, i.e., those terms that people were deemed to be unfamiliar with. For instance, the famous US series DESPERATE HOUSEWIVES is largely unknown to Germans, who represented 87% of all participants.

Though comparatively laborious, we opted for the largely automatized and randomized method presented above rather than for manual selection, which might have incurred personal bias into the design of both benchmark sets.

5.5.2 Online Survey Design

Both online studies exhibited an identical make-up, differing only in the concept pairs contained. Participants were required to rate the semantic relatedness of *all* concept pairs on a 5-point likert scale, ranging from *no proximity* to *synonymy*. 51 people completed B_1 and 23 of them also filled out B_0.[3] 87% of B_1's participants were German, while the remaining 13% were Italian, Turkish, US-American, and

---

[3] Owing to B_1's major importance, we asked people to complete B_1 first.

Israeli. 27% of all participants were CS PhD students or faculty, much lower than for Resnik's replication of Miller-Charles.

The results of each survey B_z, $z \in \{0,1\}$, were regarded as the proximity rating vector $\mathbf{v}_i \in \{1,2,\ldots,5\}^{|B_z|}$ for the respective participant i. We thus computed the *inter-subject correlation*, i.e., the Pearson correlation coefficient $p(\mathbf{v}_i, \mathbf{v}_j)$ (see Section 5.4.2.2) for every unordered pair $\{i, j\} \in B_z \times B_z$ of human subjects, represented by their proximity rating vectors. Pair similarity scores were summed up and averaged afterwards:

$$p_z = \sum\nolimits_{\{i,j\} \in (B_z \times B_z)} p(\mathbf{v}_i, \mathbf{v}_j) \cdot \frac{2}{|B_z| \cdot (|B_z| - 1)} \tag{5.7}$$

For set B_1, we obtained an average correlation $p_1 = 0.7091$. For B_0, we had an average correlation p_0 of 0.7027. These values bear strong indication for judgement correlation, but are still considerably lower than Resnik's human judgment replication of Miller-Charles, which had an inter-subject correlation of 0.8848 [Resnik, 1995].

We identify the following points as driving forces behind this observation:

- **Instances versus concepts.** Miller-Charles only featured concepts contained in dictionaries. We rather focused on names of artists, brand names, composed and qualified concepts. These are harder to handle as they are more specific and require a certain extent of domain knowledge (e.g., knowing that GMAIL is Google's electronic mail service).
- **Language and demographics.** In Resnik's replication [Resnik, 1995], ten people affiliated with CS research participated. However, in our experiment, demographics were much more wide-spread and CS researchers only had an overall share of 27% for B_1. Even larger perturbation was caused by language, for most of the participants were German-speaking while the study itself was in English.

5.5.3 Proximity Metrics

For measuring proximity, we compared several strategies which can be categorized into two larger classes, namely *taxonomy-based* (see Section 5.4) and *text-based*. While focusing on the group of taxonomy-based metrics, the second group served as an indication to verify that traditional text-based methods cannot provide better performance, thus rendering the new breed of taxonomy-based metrics obsolete for our purposes.

5.5.3.1 Taxonomy-Driven Metrics

For the taxonomy-based category, we opted for the metrics presented in Section 5.4, namely Leacock-Chodorow, Li et al, and our own approach, henceforth "Ziegler et al". For both WordNet metrics, i.e., Li et al and Leacock-Chodorow, we employed the strategy presented in Section 5.4.1 for comparing *concept lists* rather than singletons.

5.5.3.2   Text-Based Approaches

Besides taxonomy-driven metrics, we also tested various classic *text-based* similarity measures for achieving the task, based upon the well-known vector-space
paradigm [van Rijsbergen, 1975; Baeza-Yates and Ribeiro-Neto, 2011]. To this end,
we tested four setups, using two different types of data to run upon:

First, instead of using the taxonomic description $q^{c_z}(i)$ of each search query result
i for query c_z, we used its brief *textual summary*, i.e., the snippet returned by Google
to describe the respective query result. These snippets typically contain the title of
the result page and some 2-3 lines of text that summarizes the content's relevance
with respect to the query. Second, instead of using the snippet only, we downloaded
the *full document* associated with each query result i for query concept c_z.

Next, we applied Porter stemming and stop-word removal (see Baeza-Yates and
Ribeiro-Neto [2011]) to the first 100 search results (both snippets and full document) for all 260 queries crawled from Google Suggest. Both setups, i.e., snippet-
and document-based, were further subdivided into two variations each: While term
frequency (TF) [Baeza-Yates and Ribeiro-Neto, 2011] was always applied to all index term vectors, inverse document frequency (IDF) was first switched on and then
off for snippet- and document-based. Hence, we get four different setups.

5.5.4   *Experiments*

For comparing the performance across all metrics, we again followed the approach
proposed by Resnik [1995, 1999] and Li et al [2003], i.e., we computed the predicted
proximity of all concept pairs for set B_0 as well as B_1, thus obtaining the respective
metric's rating vector. Next, we computed the Pearson correlation $p(\mathbf{v}_m, \mathbf{v}_i)$ of each
metric m's rating vector $\mathbf{v}_m$ with the proximity rating vectors of all participants $i \in B_z$
for one given experiment B_z and averaged the summed coefficients:

$$p_z^m = \sum_{i \in B_z} p(\mathbf{v}_m, \mathbf{v}_i) \cdot \frac{1}{|B_z|} \qquad (5.8)$$

The correlation thus measures the metric's compliance with human ratings. The
higher the average correlation, the better.[4]

Since Li et al and Ziegler et al demand tuning parameters, we conducted two
separate runs. The first one, operating on B_0, was meant for parameter learning. The
learned optimum parameters were then used for the second run, based on B_1, i.e.,
the actual test set.

5.5.4.1   Parameter Learning

Li et al [2003] give $\alpha = 0.2$ and $\beta = 0.6$ as optimal parameters for their approach.
However, since we are supposing a different data set, we ran parameterization trials

---

[4] Opposed to Resnik [1995], the inter-subject correlation does *not* represent an upper bound
for metric correlation with human ratings, as can be shown easily.

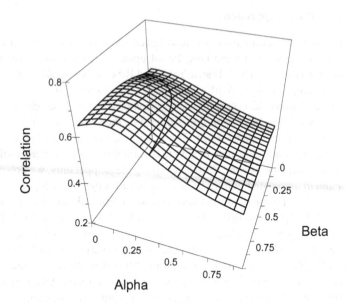

Fig. 5.2 Parameter learning curve for Li et al

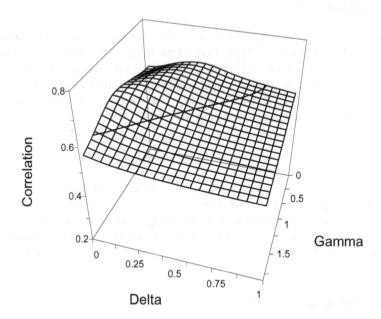

Fig. 5.3 Learning parameters γ and δ for Ziegler et al

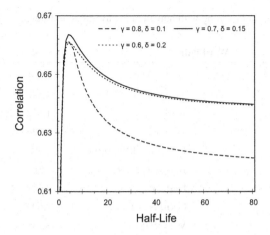

Fig. 5.4 Learning parameter half-life α in Ziegler et al

for α and β again. We thereby assumed $|q^{c_z}| = 30$ for all concepts c_z, i.e., 30 query results were considered for defining each concept/instance c_z. For both α and β, we tested the interval $[0,1]$ in .05 increments on B_0. The two-dimensional curve is shown in Figure 5.2. As optimal parameters we obtained $\alpha = 0.2$ and $\beta = 0.8$, which comes close to the values found by Li et al [2003]. The peak correlation amounts to 0.6451.

For our own approach, three parameters had to be learned, namely coefficients γ, δ and half-life α. Again, we assumed $|q^{c_z}| = 30$. Parameters γ and δ were determined first, having $\alpha = 10$, see Figure 5.3. The optimal values were $\gamma = 0.7$ and $\delta = 0.15$, giving the curve's peak correlation of 0.6609. As Figure 5.3 shows, all higher values are settled around some fictive diagonal. For probing half-life α, we therefore selected two points spanning the fictive diagonal, with the optimal parameters $\gamma = 0.7$ and $\delta = 0.15$ in the middle. The results for increasing α over all three 2D-points are shown in Figure 5.4. Again, the peak is reached for $\gamma = 0.7$ and $\delta = 0.15$, when assuming $\alpha = 7$.

The learned values were then used in the actual evaluation runs performed on B_1.

5.5.4.2 Performance Analysis

First, we evaluated the proximity prediction performance across all taxonomy-based metrics. To this end, we tested all four metrics on varying query result sizes $|q(c_z)|$ for all concepts c_z, ranging from 1 to 80 documents. Results are displayed in Figure 5.5(a), giving the average correlation with human ratings for each metric and $|q^{c_z}| \in [1,80]$. The number of topics/documents for characterizing one concept/instance appears to have little impact on Leacock-Chodorow. When $|q^{c_z}| > 40$, the correlation seems to worsen. For Li et al, an increase of $|q^{c_z}|$ has merely marginal positive effects. We owe these two observations to the fact that WordNet metrics are not designed to compare sets or lists of topics, but rather two singleton topics only

Table 5.1 Training set B_0, along with human rating averages and standard deviation

Word Pair		$\varnothing$-Rating	Std. Dev.
EMINEM	GREEN DAY	3.304	0.997
XML	VODAFONE	1.348	0.476
KING	JOBS	1.522	0.714
HOROSCOPES	QUOTES	1.826	0.761
FIREFOX	INTERNET EXPLORER	4.391	0.57
IMDB	BLOCKBUSTER	3.304	0.856
DOGS	DISNEY	2.478	0.926
FLOWERS	FOOD NETWORK	1.609	0.82
E-CARDS	VALENTINE'S DAY	3.435	0.825
FREE MUSIC DOWNLOADS	ITUNES	3.696	0.906
DELTA AIRLINES	US AIRWAYS	4.0	0.978
DELL	BEST BUY	3.13	0.947
VODAFONE	O2	4.304	0.687
MOVIES	NEWS	2.435	1.056
YAHOO MAPS	ZONE ALARM	1.696	0.953
DICTIONARY	SUPER BOWL	1.13	0.448
POEMS	LYRICS	4.087	0.83
QUICKEN	PAYPAL	2.609	0.872
NASA	KAZAA LITE	1.043	0.204
LAS VEGAS	EXCHANGE RATES	1.913	0.88
CARS	GIRLS	2.043	1.083
WEATHER CHANNEL	NEWS	3.435	0.77
GUITAR TABS	HOTMAIL	1.13	0.448
QUIZ	AMAZON	1.217	0.507
TSUNAMI	WEATHER	3.87	0.74

(see Section 5.4.1). Besides, Figure 5.5(a) also shows that Li et al has much higher correlation with human ratings than Leacock-Chodorow's simplistic metric.

For our own approach, we tested the metric's performance when using half-life $\alpha = 7$, which had been found the optimal value before, and $\alpha = \infty$. Note that an infinite half-life $\alpha = \infty$ effectively makes all topics obtain equal weight, no matter which list position they appear at. For $\alpha = 7$, the curve flattens when $|q^{c_z}| > 25$. The flattening effect appears since all topics with low ranks $i > 7$ have so little weight, less than 50% of the top position's weight. Adding more topics, considering that additional topics have increasingly worse ranks, therefore exerts marginal impact only. On the other hand, for $\alpha = \infty$, smaller fluctuations persist. This makes sense,

Table 5.2 Test set B_1, along with human rating averages and standard deviation

Word Pair		⊘-Rating	Std. Dev.
HOLIDAY INN	VALENTINE'S DAY	1.882	0.855
BLOCKBUSTER	NIKE	1.588	0.867
INLAND REVENUE	LOVE	1.216	0.604
GOOGLE	GMAIL	4.118	0.783
LOVE QUOTES	TV GUIDE	1.549	0.749
PC WORLD	UNITED AIRLINES	1.235	0.468
JOKES	QUOTES	2.294	1.108
DELTA AIRLINES	LOVE POEMS	1.098	0.357
BRITNEY SPEARS	PARIS HILTON	3.804	0.767
NASA	SUPER BOWL	1.392	0.659
PERIODIC TABLE	TOYOTA	1.176	0.55
WINZIP	ZIP CODE FINDER	1.902	1.241
EASYJET	CHEAP FLIGHTS	4.294	0.749
PEOPLE SEARCH	WHITE PAGES	3.843	1.274
TSUNAMI	HARRY POTTER	1.098	0.297
BBC	JENNIFER LOPEZ	1.686	0.779
U2	RECIPES	1.157	0.364
THESAURUS	US AIRWAYS	1.118	0.322
XBOX CHEATS	YAHOO GAMES	2.941	0.998
CURRENCY CONVERTER	EXCHANGE RATES	4.137	0.971
FLOWERS	WEATHER	2.765	1.059
USED CARS	VIRGIN	1.431	0.693
EMINEM	MUSIC	4.137	0.687
CARS	HONDA	4.176	0.617
LYRICS	REAL PLAYER	2.588	1.07
FREE GAMES	XBOX 2	2.549	0.996
MSN MESSENGER	UPS	1.706	0.799
MICROSOFT	INTERNET EXPLORER	4.314	0.727
POEMS	SONG LYRICS	3.647	0.762
DELTA AIRLINES	WALMART	1.725	0.794

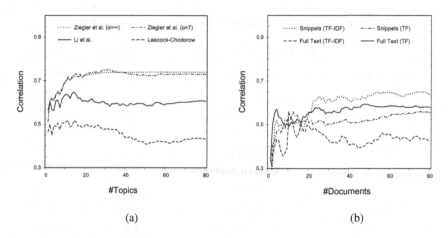

Fig. 5.5 Correlations of the three proximity metrics with human ratings, for increasing number of topics (a) and documents (b)

for every added topic has *equal* impact. However, the curves for $\alpha = 7$ and $\alpha = \infty$ exhibit differences of smaller extent only. When assuming more than 40 topics per concept, correlation worsens somewhat, indicated through the $\alpha = \infty$ curve. As opposed to both WordNet metrics, our metric performs better when offered more information, i.e., more topics per concept. The latter finding backs our design goal geared towards *multi-class* categorization (see Section 5.4.2).

Moreover, Figure 5.5(a) shows that Ziegler et al performs *significantly* better than both other taxonomy-based benchmarks. With $\alpha = \infty$, the peak correlation of 0.7505 is reached for $|q^{cz}| = 31$. Curve $\alpha = 7$ levels out around $|q^{cz}| = 30$, giving a correlation of 0.7382. For comparison, Li et al's peak value amounts to 0.6479, Leacock-Chodorow's maximum lies at 0.5154.

Next, we compared the performance of *text*-based proximity metrics, shown in Figure 5.5(b). All metrics, except for full text-based with TF and IDF, drastically improve when offered more documents for representing one concept. However, fluctuations are much stronger than for the taxonomy-based metrics. For more than 20 documents per concept, snippets-based with TF and IDF performs best, reaching its maximum correlation of 0.6510 for 73 documents. This performance is comparable to Li et al's, but while the text-based metric becomes more accurate for document numbers > 20, the mentioned taxonomy-based metric exhibits better performance for topic numbers < 20.

5.5.4.3 Conclusion

We have shown that our novel metric outperforms both state-of-the-art taxonomy-based proximity metrics as well as typical text-based approaches. For reasonably large numbers of topics , i.e., $|q^{c_z}| > 20$, the correlation with human ratings lies between 0.72 and 0.75, indicating strong correlation. Leacock-Chodorow, being an utterly simplistic taxonomy-based approach, and the full text-based approach with TF *and* IDF, both exhibited correlations below 0.5 for more than 20 topics/documents. Opposed to our approach, their performance was better when using *less* information, i.e., < 20 topics/documents, still peaking only slightly above 0.5. The other three metrics, i.e., Li et al, snippet-based with and without IDF, and full text-based without IDF, had correlation scores between 0.55 and 0.65 for more than 20 topics/documents.

5.6 Outlook

Semantic proximity metrics are becoming increasingly important in ecosystems that are characterized by massive amounts of textual information, such as the Web 2.0. Currently, only for small portions of information fragments, namely words and simple concepts stored in thesauri and dictionaries such as WordNet, semantic similarity measures are applicable. By harnessing the combined power of both Google and ODP, we were able to extend semantic proximity to *arbitrary* concepts (called named entities), e.g., names of persons, composed concepts, song titles, and so forth. Moreover, we introduced a new taxonomy-based proximity metric that showed significantly better performance than existing state-of-the-art approaches and comes close to human judgement.

The framework for calculating the semantic proximity between two arbitrary concepts *does* have an application scenario in recommender systems. In particular for content-based approaches, where user-item or item-item similarities are calculated based on *descriptive features* rather than *behavior*, we can put the semantic proximity concept to effective use: For more of often than not, these product descriptions are given by *textual* rather than *structured* information.

Chapter 6
Recommending Technology Synergies

" The whole is greater than the sum of its parts."

– Aristotle (384 BC – 322 BC)

6.1 Introduction

Chapter 5 has laid out an approach to identify the semantic proximity between two named entities like brand and product names, locations, and so forth. The underlying chapter will now work on this model and put it to use for the identification and recommending of technology synergies. That is, given a set of technologies associated with different business units, we want to know which are the pairs of technologies that exhibit the largest synergies.

This recommender of technology synergies is different from the ones we have seen before in two ways: First, it is not operating on recommending *products* based on purchasing behavior. Second, it is *non*-personalized and thus may not count as a pure recommender system in the strict sense. It resembles the ones seen before by making abundant use of Web 2.0 information, namely taxonomies and the collective wisdom of Wikipedia (*http://en.wikipedia.org*).

6.2 Motivation for Recommending Technology Synergies

Large corporations often operate in diverse business segments across different industries. While these various industry branches (e.g., health-care solutions, electrical power generation, or industrial automation) may appear to have little overlap with regard to their nature and products, they commonly *do* have significant overlap at a more basic level, namely the level of core or support technologies employed.[1] For instance, CIRCUIT BREAKERS as well as SWITCH GEAR devices are

---

[1] The term "technology" is used in a broader sense in this paper, meaning any craft that expresses progress of any kind (not necessarily IT-related) of human civilization.

C.-N. Ziegler: *Social Web Artifacts for Boosting Recommenders*, SCI 487, pp. 79–95.
DOI: 10.1007/978-3-319-00527-0_6   © Springer International Publishing Switzerland 2013

designed and used in departments subordinate to different operating units and different branches in a company, e.g., industrial automation as well as the energy sector.

To adopt a more competitive market position and allow for significant cost savings, there is an increasing interest in consolidating technologies across operational units and in exploiting existing synergies that have not been discovered before. Within the strategic planning process in a company, operating units usually describe their basic technologies, the ones already in use as well as those currently being developed. The resulting technology lists are then matched against each other, comparing pairs of technologies $\{t_i, t_j\}$ on the basic level for possible synergies and overlap. While this task appears trivial when both compared technologies t_i, t_j bear identical names or when semantic relatedness is reflected in lexical similarity, it becomes complex when these assumptions do not hold anymore, which is generally the case. For instance, WASTEWATER TREATMENT and FORWARD OSMOSIS are related technologies with relatively high overlap, even though they are not lexically similar. However, both are about HYDRAULIC PERMEABILITY and FILTRATION and, in addition, based on similar concepts and mechanics.

At an integrated technology company like Siemens[2], the evaluation of synergies has traditionally been done in a manual evaluation process by various technical domain experts. Next to being tremendously expensive in terms of resources, the overall duration for completing these reports has been very high. We have addressed this issue by designing and deploying a service that is able to recommend technological synergies for Siemens operating units in an *automated* fashion, by applying Web mining and information retrieval techniques Baeza-Yates and Ribeiro-Neto [1999] that work on data gathered from the Web 2.0's collective intelligence.

Our system's output is twofold: First, there is an $N \times N$ pivot table[3] that gives the extent of overlap on different organizational levels; e.g., allowing us to deduce that sector s_1 has more overlap with sector s_2 than s_3. Second, there is a ranking module that returns the list of pairs in descending order according to their estimated synergetic potential. Next to the ranking itself, the system also delivers an explanation detailing *why* the technology pair at hand has a high synergy value.

In order to show our system's accuracy, we have conducted empirical evaluations, based on random samples of pairs of technologies $\{t_i, t_j\}$; these pairs were rated for perceived synergies by two control groups, one being a group of general technologists with a background in computer science, one being a group of experts that used to evaluate technology pairs by hand in the past. Our automated approach's result was then matched against the two groups' consensus, showing that our system performs considerably well and is virtually indistinguishable from a man's assessment in terms of classification accuracy.

The synergy discovery system is in operative use and saves considerable human effort as we speak, enabling experts to spend their valuable time within the synergy exploitation process more efficiently. Its usage has been likewise embraced by

---

[2] Note that the research presented in this chapter was conducted during the author's tenure at Siemens Corporate Research and Technologies.

[3] Where N is the overall number of technologies being compared.

domain experts and management and counts as paradigmatic example on how to make effective use of the Web 2.0's intrinsic value for enterprises.

6.3 Scenario Description

Synergies are identified on the purely technological level, estimating synergy values for two given technologies t_i, t_j, as well as on aggregated organizational levels. To this end, two levels are considered, namely the top-most organization level of three sectors (industry, energy, and health-care), as well as the subdivision of sectors into 15 divisions. For each division, d_i, a list of 10 representative technologies is composed. The complete technology list therefore is comprised of 15×10 entries, where each entry is a triple $(t_i, s_j, d_k) \in T \times S \times D$, with T denoting the technologies, S giving the sectors, and D the divisions. Table 6.1 provides an example excerpt for one such list. Assuming that each technology occurs once, the number of technology pairs to be checked is as follows:

$$\left| \left\{ \{t_i, t_j\} \in T \times T \mid i \neq j \right\} \right| = \frac{150 \times 149}{2} = 11,175 \qquad (6.1)$$

These roughly 11,000 pairs have been inspected annually by domain experts in a manual fashion so far, requiring the effort of several person months of skilled experts. Since our system's deployment, these computations take less than one day and help us to reduce the human effort dramatically to some 5-10%, by virtue of weeding out apparently non-synergetic pairs.

Table 6.1 Sample technology list

Technology	Sector	Division
COMPUTED TOMOGRAPHY	Health-Care	Div A
SMOKE DETECTOR	Industry	Div B
FLUE GAS DESULFURIZATION	Energy	Div C
MEMBRANE BIO REACTOR	Industry	Div D

...

6.4 System Setup

Our system computes scores in an automated fashion by relying on two classifiers. The first one makes effective use of Wikipedia (*http://en.wikipedia.org*) as background knowledge source, the second one relies on ODP, the Open Directory Project (*http://www.dmoz.org*). Both sources are maintained and continually updated by several thousands to millions of voluntary users who collectively contribute to the creation of the world's largest knowledge repositories.

In order to come up with one single synergy score for each pair $\{t_i, t_j\}$, the two classifier scores must be merged appropriately. However, as the distribution of weights of both classifiers deviates drastically from each other, we decided to rank pairs along each classifier and merge the two ranks into one. An additional confidence score indicates whether the two assigned classifier ranks reach consensus or not.

The two classifiers, denoted C_W for Wikipedia and C_D for ODP, operate in a different fashion and focus on different aspects of what makes two technologies appear synergetic: C_W tends to identify synergies by detecting primarily *merony-mous* relationships in both technologies, while C_D concentrates more on *taxonomic* relationships between both. For instance, the synergy between the aforementioned pair WASTEWATER TREATMENT and FORWARD OSMOSIS can be characterized by referring to components and processes they have in common, i.e., HYDRAULIC PERMEABILITY and FILTRATION; the focus is therefore on part-whole (i.e., meronymous) relationships. On the other hand, the synergy can also be quantified by assigning both technologies to *classes* within a classification scheme and look-ing for the least common subsumer. For the case at hand, the subsuming category could be WASTEWATER TECHNOLOGIES, for both technologies are specializations of that very category.

The following two sections describe the classifiers in detail.

6.4.1 Wikipedia-Based Classifier

Wikipedia is a massive information source featuring more than 2.3 million articles at the time of writing. In order to compute the synergy score for two technologies t_i, t_j, identified by their name, our system first downloads the respective articles from Wikipedia, denoted $p(t_i)$ and $p(t_j)$. These HTML documents can be accessed by using the site's internal search service. We only allow exact matches, which puts some constraints on the input data being used.

Next, we extract all hyperlinks contained in $p(t_i)$ and $p(t_j)$, resulting in two link sets $L_p(t_i)$ and $L_p(t_j)$. Only links internal to Wikipedia are considered, i.e., those links that refer to other articles in Wikipedia. Moreover, non-content related links are discarded from these two sets. Those comprise of links to disclaimer information, terms of use, as well as hyperlinks to geographical locations, i.e., cities, countries, and so forth. Our investigations have shown that such links are of marginal use for the content's description and rather add noise to the classification process. Non-content related links are held in a black list that we have created manually by probing and analyzing sample sets before deployment.

Eventually, the classifier C_W computes the synergy score $s_W(t_i, t_j) \in [0, 1]$ as fol-lows, where $L'_p(t_i)$ and $L'_p(t_j)$ denote the hyperlink sets after removing irrelevant links:

$$s_W(t_i, t_j) := \frac{\left| L'_p(t_i) \cap L'_p(t_j) \right|}{\left| L'_p(t_i) \cup L'_p(t_j) \right|} \qquad (6.2)$$

The number of common links is normalized by means of dividing by the union of links in both sets.

As has been said before, each classifier also generates a description that indicates *why* the pair at hand is assigned a high value. For C_W, this is simply the link text that comes with the hyperlinks both downloaded documents have in common. For instance, for the technology pair of COMBUSTION and ALTERNATIVE FUEL, we have FUELS, HYDROCARBONS, INTERNAL COMBUSTION ENGINES, etc.

The distribution of score for the complete set of selected technology pairs is displayed in Fig. 6.1 and will be discussed in detail later.

6.4.2 ODP-Based Classifier

The second classifier, C_D, is substantially more complex than the comparatively simplistic Wikipedia-based one. Its inner mechanics are described in depth in Chapter 5 and [Ziegler et al, 2006].

As we know from the preceding chapter, C_D depends on various parameters, namely half-life α and two other parameters that have an impact on the propagation of score from leaf nodes to the root. These parameters have been learned on an external test set of terms which were not included for the classifier's later application to "real" technology pairs. To this end, we have composed a small list of primarily IT technologies (see the "list of IT management topics" on Wikipedia), had an expert rate the synergies of all pairs, and applied non-linear optimization techniques so as to find out optimal values for all three parameters of C_D.

Classifier C_W returns, next to the synergy estimate, an explanation for the recommended score (see Section 6.4.1). For C_D, this is also the case: The ODP-based classifier also returns an ordered list of the 20 ODP categories that have been assigned the highest scores both for $L(t_i)$ and $L(t_j)$. As a rule of thumb we can say that these 20 categories stem from higher taxonomy levels when the estimated synergy score is low, and come from lower (i.e., more specific) levels when the synergy score is high.

6.4.3 Merging Classifiers

In order to come up with one single classification score for pairs $\{t_i, t_j\}$, the results of C_W and C_D need to be merged into one. However, Figure 6.1 shows that the distribution of score weight for both classifiers is substantially different[4]: Classifier C_D appears to distribute score in a more homogeneous fashion over the full target value range $[0, 1]$, while C_W assigns higher scores to much fewer pairs. Adding up C_W's and C_D's score is therefore not an appropriate option for score merging, as

---

[4] The distributions have been computed based on the complete set of $11,175$ technology pairs.

C_W's impact would be comparatively low. Scaling C_W would likewise fail to address the issue.

We therefore opted for another approach, namely to order all pairs $\{t_i, t_j\}$ according to their *rank*, based on s_W and s_D, respectively. Thus, we introduce two ranking maps $R_W : T \times T \to \mathbb{N}$, R_D defined accordingly. So as to merge both ranks into one, we take the average of both, resulting in function $s(t_i, t_j) := (R_W(\{t_i, t_j\}) + R_D(\{t_i, t_j\}))/2$ which represents the base for computing the final rank R. The drawback of that approach is that relative differences from pairs are not necessarily preserved, owing to equidepth binning.

In order to verify whether the two classifiers were reaching consensus with regard to the classification task, we introduced a simple rating consistency metric $c : T \times T \to [0, 1]$ that would tell for each pair whether C_W and C_D do comply or not:

$$c(t_i, t_j) := 1 - \frac{|R_W(\{t_i, t_j\}) - R_D(\{t_i, t_j\})|}{|T| \times (|T| - 1) \times 0.5} \tag{6.3}$$

When C_W and C_D assign the same rank, function c returns 1, attesting 100% consensus. When they dissent maximally, the result is 0. Figure 6.2 shows averaged consensus over all pairs. The mean consensus is .71, which hints at largely compliant classifiers. Interestingly, consensus is highest for pairs ranked at positions 4,000 to 6,000. It is lowest for the bottom ranks beyond 10,000. Inspection of the data showed that toward the list's bottom, large blocks of pairs with very low ranks R_D and moderate ranks R_W appeared. However, an examination of samples gave no indication for misclassification; i.e., virtually no false negatives (true synergies that were predicted as "non-synergetic") were discovered.

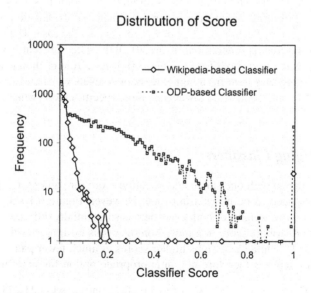

Fig. 6.1 Score distributions for C_W and C_D

6.5 System Usage

Our system takes as input one single list that contains triples (t_i, s_j, d_k) made up of technology t_i, sector s_j, and division d_k. However, before the list becomes digestible for the automated classification, the *findability* of these t_i needs to be verified. That is, for each technology, we must make sure that there is a corresponding article on Wikipedia as well as a search result on Google Directory.

From the original list of 150 selected technologies, only 20% could be found on Wikipedia when searching for the literal text. At the same time, Google Directory search results comprising of more than 5 entries were found for circa 60% of the selected technologies. Recall that we made use of *phrasal* search queries only, so as to attain higher accuracy. That is, all technologies consisting of more than one word, e.g., ARC-FAULT CIRCUIT INTERRUPTER, were put into quotations "" in order to have the search engine treat them as one coherent text token. Without query phrasing, results would have been returned for virtually all queries.

6.5.1 Pre-processing

For list pre-processing, the domain experts' task was thus to find technologies t_i' that were retrievable both on Wikipedia and Google Directory *and* were semantically narrow to the original technology t_i. Either replacing t_i's name by its syntactically

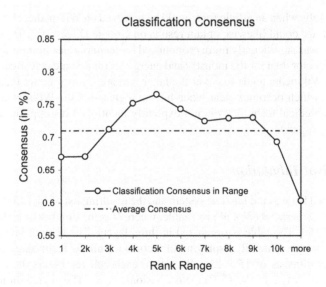

Fig. 6.2 Average consensus over rank range

	Industry	Energy	Health	
	56.53	58.09 [(I)]	36.46	Industry
		65.38 [(E)]	39.43	Energy
			48.12 [(H)]	Health

Fig. 6.3 Inter- and intra-sector synergy averages

different but semantically equivalent[5] concept label or, if not feasible, superseding t_i by its slightly more general super-concept t'_i appeared as the most obvious approach. However, the use of similar technologies (rather than the original ones) incurs some risk, namely to add noise to the detection of synergies and thus to deteriorate the result.

Interestingly, when an article for t_i could be retrieved on Wikipedia, chances were around 95% we could likewise obtain results on Google Directory. The noted findability issue was significantly more pronounced for technologies stemming from the health-care sector than for the industry and energy sectors. Some investigations have shown that Wikipedia tends to cover the latter two areas much better than medical technology, which becomes clear when having a glance at the Wikipedia category tree, where medical technology is not explicitly mentioned (as opposed to energy and industry topics).

6.5.2 Pivot Tabulation

The processed list was fed into our system and the resulting list of 11,175 technology pairs used to generate an $N \times N$ pivot table, where N is the number of technologies, in our case 150. The table is organized in three layers: The bottom layer features the highest degree of detail, displaying the scores for all the technology pairs. The second layer consists of 15×15 cells, where each cell represents the compound synergy score for two of the 15 Siemens divisions, d_x, d_y. The aggregation function used for computing the cell value of d_x and d_y is the average of scores of technology

---

[5] As in virtually every company, certain concepts at Siemens have names that are used internally only.

pairs $\{t_i, t_j\}$, where t_i is one of the technologies assigned to d_x, and t_j is one of the technologies assigned to d_y, respectively. For calculating the average division-internal score, i.e., the synergy score for technologies within one given division $d_x = d_y$, we take all pairs $\{t_i, t_j\}$ except those where $i = j$.

The third and most coarse-grained layer measures the synergetic potential of the selected technologies across and within sectors. Again, the aggregation is computed in the same fashion as for the divisions.

For our pivot tabulation, we use the merged ranking R of all the technology pairs $\{t_i, t_j\}$ as basis for score computation. Hereby, we subdivide the full rank range into 100 equiwidth bins. A middle-ranked technology pair thus gets score value 50.

6.5.3 Hypothesis Validation

Next to providing an integrated overview of all likely technology synergies within the organization on different levels of granularity, the pivot tabulation allows us to easily validate the system's classification ability by means of some simple hypotheses. With regard to the sectors, an intuitive hypothesis is that the *intra*-sector technology overlap, i.e., the synergetic potential per se, is greater than the *inter*-sector overlap. When using an automated classification scheme as ours, we would expect the classifier to likewise reflect this hypothesis.

Figure 6.3 shows the average synergy potential for all combinations of sectors, likewise comprising of inter- and intra-sector scores; higher cell values indicate higher synergetic potential. Hereby, the sector pair which exhibits the highest average synergy score for one given sector is decorated with one of three superscripts (I), (E), and (H), being acronyms for the respective sectors. We can thus observe that our hypothesis of higher intra-sector scores as opposed to inter-sector scores holds for two of the sectors: For the health as well as the energy sector their intra-sector score is significantly higher than their respective inter-sector scores. This is not the case for the industry sector, though, where the machine suggests a higher synergetic potential for the cross-sector overlap with energy than within its own sector. However, the scores are close to each other (58.09 versus 56.53). Moreover, it is an established fact that the energy and industry sector are in general much closer to each other than the health sector, which is also reflected in the data.

Table 6.2 Averages of inter- and intra-division scores

Sector	Intra-Division	Inter-Division
Industry	62.66	59.71
Energy	64.70	63.32
Health-Care	54.75	28.45

The same effect can be observed on the finer-grained division level. Here, the increase in overlap of technologies *within* divisions as opposed to *across* divisions is even more pronounced than for the sector level. Table 6.2 lists the average inter-division synergy potential versus the average intra-division synergy score for each sector. Notably, the difference is enormously large for the health-care sector which indicates that the various divisions deal with substantially different topics and are well-separated from each other.

Eventually, we may come to conclude that our intuition of coherently arranged organizational units is reflected in our system's automatic synergy prediction approach, even though the results are based on representative selections rather than the full range of technologies.

6.6 Synergy Classifier Evaluation

Our classifier system attempts to judge the synergetic potential of any two technologies t_i, t_j and computes an explicit ranking R of technology pairs according to their synergy value, enabling us to come up with a synergy recommendation. The previous section has shown that the use of the classifier in hierarchical pivot tabulating according to organizational layers reveals several findings we would have intuitively expected, thus serving as means of validation.

However, in order to properly quantify the classification accuracy of the system, benchmark evaluations are required. Owing to the lack of relevant labeled datasets that could be used for cross-validation, the only reasonable option is to have humans judge a subsample of the overall set of technology pairs and to compare the machine's automatically computed result with human ratings.

6.6.1 Experiment Setup

We have composed one such benchmark test set made up of a sub-sample of all technology pairs. To this end, we have subdivided the 11,175 pairs, sorted according to R, into four bins and randomly selected 5 pairs from each bin. Each pair is assigned its bin number, which simultaneously expresses the synergy value on a 4-point scale. The selected pairs are shown in Table 6.4.

Next, we had two groups of people, consisting of 10 members each: The first was made up of computer science researchers and IT consultants with an affinity to technology in general but no in-depth knowledge of non-IT production processes and products. This group was considered the *non-expert group*. The second cluster, dubbed the *expert group*, only contained people with relevant domain expertise. However, though the latter group is named "expert group", not all its members are experts for all three sectors at the same time. Most of them specialize in merely one of them (or even a subset of one, by only focusing on some of the divisions of one given sector).

The 20 technology pairs were shuffled, in order to discard the machine's ranking from top-synergy pairs down to those "poor dogs". Next, the list was shipped to all

Table 6.3 Benchmark set with machine ratings and ratings by experts and non-experts (I/II)

Word Pair		Machine		Experts		Non-Experts	
		Rating	$\oslash$-Rating	$\oslash$-Rating	Std. Dev.	$\oslash$-Rating	Std. Dev.
PHOTOVOLTAICS	ALTERNATIVE FUEL	4	1.8	1.8	0.63	3.4	0.7
DIRECT DRIVE MECHANICS	PICTURE ARCHIVING SYSTEM	1	1.0	1.0	0.0	1.4	0.7
HEAT RECOVERY	SOLAR THERMAL ENERGY	3	2.4	2.4	1.26	3.2	0.79
SMOKE DETECTOR	STEAM TURBINE	2	1.6	1.6	0.97	1.8	0.92
REMOTE MONITORING	ARC-FAULT CIRCUIT INTERUPTER	3	3.2	3.2	0.79	3.2	0.79
TELEMATICS	CONTROL SYSTEM	2	2.9	2.9	1.10	3.7	0.48
SILICON CARBIDE	COLD ROLLING	3	1.1	1.1	0.32	1.9	0.32
PC-BASED AUTOMATION	LABORATORY AUTOMATION	2	3.2	3.2	1.03	3	1.05
CARBON CAPTURE	ALTERNATIVE FUEL	4	2.0	2.0	0.67	2.9	0.99
COMPUTED TOMOGRAPHY	CERAMICS	2	1.6	1.6	0.70	1.5	0.71

Table 6.4 Benchmark set with machine ratings and ratings by experts and non-experts (II/II)

Word Pair		Machine	Experts		Non-Experts	
		Rating	Ø-Rating	Std. Dev.	Ø-Rating	Std. Dev.
MAMMOGRAPHY	COMPUTED TOMOGRAPHY	4	3.2	0.79	3.2	0.63
AIRPORT	VIRTUAL ENGINEERING	1	2.2	1.03	2.2	1.14
C-ARM	MEDICAL IMAGING	3	1.6	0.84	1.8	0.63
ROLL FORMING	FOIL BEARING	1	1.4	0.52	2.2	0.79
FIRE-PROTECTION	DUAL-ENERGY SCAN	1	1.0	0.0	1.1	0.32
SWITCH GEAR	CIRCUIT BREAKER	4	3.0	1.15	2.9	1.1
GAS TURBINE	STEAM TURBINE	4	2.9	0.99	3.6	0.52
ENERGY EFFICIENCY	LED LAMP	2	2.8	0.92	2.7	0.95
SENSOR	CARBON NANOTUBE	3	1.6	0.70	2.1	0.74
RADIOPHARMACOLOGY	RADAR	1	1.0	0.0	1.6	0.7

20 participants without them knowing either the machine's rating or their fellows' estimate. The only directive given to them was to rate the synergetic value of all pairs on a scale of 1 to 4, where 1 denotes no synergy at all and 4 denotes maximum synergetic potential. Moreover, they were not forced to stratify their ratings, i.e., to assign the same number of 1's, 2's, 3's, and 4's to the list of 20. Experts as well as non-experts were allowed to look up technologies and their meanings in whatever resources they deemed useful.

6.6.2 Result Analysis

In order to analyze and compare the results, we made use of the popular Pearson correlation coefficient (see Chapter 2): The ratings of each participant, i.e., experts, non-experts, and the automated computation scheme, are considered as *vectors* where each component may adopt values between 1 and 4. The correlation coefficient is then computed for two of these vectors, returning values in the range [-1,+1].

The advantage of Pearson correlation, as opposed to for example the cosine similarity measure [Baeza-Yates and Ribeiro-Neto, 2011], lies in its taking care of the general rating tendency of the two arbiters involved. Some people rather assign higher scores while others tend to assign lower values. Pearson only measures, for each individual, her deviation from her mean rating. These deviations from mean ratings are then compared for each vector component, that is, for each technology pair being evaluated with regard to synergetic potential.

6.6.2.1 Intra-group Correlation

First, we wanted to analyze how members of each of the two groups would behave internally with regard to their synergy judgment. We therefore computed the correlation coefficient for every unordered pair of members $\{m_i, m_j\}$ of each group, discounting self-correlations $i = j$. The result was then averaged across the number of pairs, $|\{\{m_i, m_j\}, i \neq j \wedge i, j \in \{1, 2, \ldots, 10\}\}| = 9 \cdot 10/2$.

For the non-expert group, the average intra-group correlation amounted to .53, which is a fairly high correlation. The fact that the correlation does not reach into higher ranges already indicates that synergy identification is a non-trivial task where humans likewise dissent with each other. The intra-group correlation for the expert group was .56, which supports our hypothesis that experts are closer to each other with regard to their estimation, owing to their familiarity with the process and the technologies discussed. However, the noted difference in average correlation is relatively small. Figure 6.4 provides more insight by exposing the distribution of correlation values for each member pair and group: We observe that there are more pairs for the expert group that have high correlations than there are for the non-experts. However, the expert group also has more pairs that have considerably low correlations $< .3$.

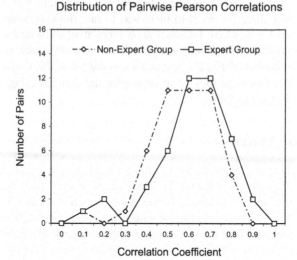

Fig. 6.4 Comparison of group distributions

6.6.2.2 Inter-group Correlation

Next, we wanted to investigate whether experts and non-experts substantially differed from each other with regard to their estimates. We thus computed the inter-group correlation that we defined as the average of correlation coefficient values for pairs $\{m_i, m_k\}$, where m_i is an expert and m_k is not. The resulting correlation of .49 hints at the finding that the *within*-group coherence is greater than consensus *across* both groups. That is, experts behave in a certain fashion and non-experts do so as well. However, the figures are too close to each other, with no statistical significance for $p < .05$ using Student's two-tailed t-test. More extensive studies on a larger scale would be required in order to support or reject the respective assumption.

6.6.2.3 Classifier Benchmarking

Upon measuring the behavior of humans when confronted with the task, we eventually evaluated our approach's classification results by means of comparing with the human groups' assessments. We thus computed the average correlation of the automatically assigned synergy ratings with the synergy weights of all members of each group. Moreover, we also generated a non-stratified random synergy rating vector in order to define the borderline, and likewise matched the vector against both groups.

The random rating vector's fit with the non-expert group amounted to .12, and to .17 with the expert group. Our automated classification scheme's fit with the non-expert group exceeded this borderline figure by far, returning .5 as average correlation. This correlation score is almost as high as the non-experts' intra-group correlation coefficient and gives us an indication that our automated computation system's performance effectively meets our expectations. The fit with our expert group was lower, though, amounting to .46.

The results for the benchmark as well as the inter-group and inter-group correlations of both human groups are summarized in Fig. 6.5.

6.6.3 Conclusion

The empirical evaluations have shown that evaluating the synergetic potential is not an easy task for humans either. The level of expertise plays an important role, though, as domain experts appear to more strongly comply with each other than non-experts. However, the difference in inner-group consensus for these two groups was still relatively small. One reason might be that our "non-experts" effectively *did* have technical backgrounds, keeping their spread of ratings at bay. Moreover, both groups were invited to look up technologies they were not familiar with.

Of particular interest is clearly the performance of our automated synergy recommender system, which performed utterly well against the human-set benchmark. In particular for the non-expert group the fit of our classifier appeared high, which would make its result barely distinguishable from a man's assessment. While the correlation of our classifier was lower for the expert group, the fit was still comparatively good.

Hence, we come to find that our system is well-suited for use as surrogate for tedious human effort. While this assumption was already made during numerous

Experts	Non-Experts	Classifier	Random	
0.56	0.49	0.46	0.17	Experts
	0.53	0.5	0.12	Non-Experts
		n/a	0.14	Classifier
			n/a	Random

Fig. 6.5 Empirical benchmark results

on-the-fly trials throughout the system's deployment phase and during arbitrary probing, the preceding empirical evaluations have confirmed this hypothesis.

6.7   Related Work

To the best of our knowledge, there exists no system that addresses the automatic recommendation of technological synergies per se. However, as our idea of leveraging technological synergies is intrinsically connected to the notion of semantic similarity, literature on the latter subject appears most relevant for positioning our work into current research: Logics-based approaches to deriving semantic similarity come first and foremost from the realm of description logics, which are formalisms used to describe concepts and their characteristic features. Seminal works include Borgida et al [2005] and Lutz et al [2003]. The drawback of these methods is the need to concisely formalize all knowledge used throughout the decision process, which would in our case offset the advantage of machine-supported synergy detection as opposed to manual examination by dedicated domain experts.

Other related work, inspired by statistics and probabilistic reasoning, intends to determine the semantic distance of *word* meanings rather than that of generic *concepts*. To this end, these approaches mostly rely on electronically available, hierarchically structured dictionaries such as WordNet (*http://wordnet.princeton.edu/*) [Miller, 1995]. The problem of similarity matching is here reduced to simple graph traversal tasks [Budanitsky and Hirst, 2000] as well as finding least common subsumers in taxonomy trees (e.g., [Li et al, 2003], [Ganesan et al, 2003], and [Maguitman et al, 2005]). In a similar vein, researchers have proposed information-theoretic approaches, see Resnik [1999] and Lin [1998]. Unfortunately, these approaches mostly apply to concepts found in *dictionaries* and not arbitrary named entities, e.g., HYDRAULIC PERMEABILITY.

With the advent and the proliferation of the Web, corpora-centric approaches have gained momentum: These efforts compute the semantic similarity between two given words or named entities based on massive document collections (such as the Web) and use language statistics such as point-wise mutual information (see, e.g., [Turney, 2001]), word collocation or occurrence correlation [Bollegala et al, 2007] in order to estimate their semantic similarity. More complex examples include Ziegler et al [2006] and Sahami and Heilman [2006].

While these approaches appear related to our computation scheme in terms of some of the principle methods used, none of them addresses the context of deriving the synergetic potential based on meronymous and class-subclass relationships of technologies.

6.8   Discussion

In this chapter we have laid out our systematic approach for recommending technology synergies in an automated fashion. The major advantage of this method is the tremendous reduction of human effort while still returning results that match those of human experts with regard to classification accuracy (see Section 6.6.2.3).

We harness the system in two ways: First, we make use of it in order to generate multi-level $N \times N$ pivot tables that show the company's synergy potential landscape at one glance. Drill-down and roll-up operations along organizational hierarchies (sections, divisions) allow to refine or coarsen the maintained perspective (see Section 6.5.2). Second, we use the system as some sort of "synergy recommender", where we have our software rank technology pairs in descending order with regard to their synergetic value and then select the top 5-10% for further in-depth inspection. The filtered results are then offered to domain-specific working groups that decide which synergy recommendations are worthwhile to pursue and which are not.

For future work, we intend to try classifiers other than the ODP- and Wikipedia-based ones presented here. Moreover, we plan to include more rigid evaluation of classifiers competing against each other. So far, we have only conducted informal assessments for comparing the accuracy of the combination of both classifiers, as opposed to the use of just one of them. When having several additional classifiers in place, we envision to apply boosting (such as AdaBoost.M1 [Freund and Schapire, 1996]) and bagging techniques for optimal classifier weighting.

Part III
Social Ties and Trust

Chapter 7
Trust Propagation Models

"To be trusted is a greater compliment than being loved."

– George MacDonald (1824 – 1905)

7.1 Introduction

Part II has discussed making use of taxonomies for improving recommender systems, in particular with regard to the *quality* of recommendations. Part III will now shift the focus from taxonomies to interpersonal trust, which abundantly manifests via the rife social networks and platforms on the Web 2.0. In contrast to taxonomies, trust is a means to address the scalability and cold-start problem of recommenders.

The chapter at hand focuses on *trust metrics*, i.e., network-based tools for predicting the extent of interpersonal trust shared between two human subjects. Though not directly related to recommender systems research, the contributions made therein are of utter relevance for their later integration into the decentralized recommender framework.

7.2 Contributions

The main contributions of this chapter are the following two:

Trust metric classification scheme. We analyze existing trust metrics and classify them according to three non-orthogonal features axes. Advantages and drawbacks with respect to decentralized scenarios are discussed and we formulate an advocacy for local group trust metrics.

Appleseed trust metric. Compelling in its simplicity, our Appleseed local group trust metric borrows many ideas from spreading activation models [Quillian, 1968], taken from cognitive psychology, and relates their concepts to trust evaluation in an intuitive fashion. Moreover, extensions are provided that make our trust metric handle *distrust* statements, likewise.

C.-N. Ziegler: *Social Web Artifacts for Boosting Recommenders*, SCI 487, pp. 99–131.
DOI: 10.1007/978-3-319-00527-0_7   © Springer International Publishing Switzerland 2013

7.2.1 On Trust and Trust Propagation

In our world of information overload and global connectivity leveraged through the
Web and other types of media, social trust [McKnight and Chervany, 1996] between
individuals becomes an invaluable and precious good. Trust exerts an enormous impact on decisions whether to believe or disbelieve information asserted by other
peers. Belief should only be accorded to statements from people we deem trustworthy. Hence, trust assumes the role of an instrument for "complexity reduction"
[Luhmann, 1979]. However, when supposing huge networks such as the Semantic
Web, trust judgements based on personal experience and acquaintanceship become
unfeasible. In general, we accord trust, defined by Mui et al [2002] as the "subjective expectation an agent has about another's future behavior based on the history of
their encounters", to only small numbers of people. These people, again, trust another limited set of people, and so forth. The network structure emanating from our
person (see Figure 7.1), composed of trust statements linking individuals, constitutes the basis for trusting people we do not know personally. Playing an important
role for the conception of decentralized infrastructures, e.g., the Semantic Web, the
latter structure has been dubbed the "Web of Trust" [Golbeck et al, 2003].

Its effectiveness has been underpinned through empirical evidence from social
psychology and sociology, indicating that *transitivity* is an important characteristic
of social networks [Holland and Leinhardt, 1972; Rapoport, 1963]. To the extent that
communication between individuals becomes motivated through positive affect, the
drive towards transitivity can also be explained in terms of Heider's famous "balance
theory" [Heider, 1958], i.e., individuals are more prone to interact with friends of
friends than unknown peers.

Adopting the most simple policy of trust propagation, all those people who are
trusted by persons we trust are considered likewise trustworthy. Trust would thus
propagate through the network and become accorded whenever two individuals can
reach each other via at least one trust path. However, owing to certain implications
of interpersonal trust, e.g., attack-resistance, trust decay, etc., more complex metrics
are needed to sensibly evaluate social trust. Subtle social and psychological aspects
must be taken into account and specific criteria of computability and scalability
satisfied.

In this chapter, we aim at designing one such complex trust metric[1], particularly
tailored to social filtering tasks (see Section 2.3) by virtue of its ability to infer
continuous trust values through fixpoint iteration, rendering ordered trust-rank lists
feasible. Before developing our trust metric model, we analyze existing approaches
and arrange them into a new classification scheme.

---

[1] Note that trust concepts commonly adopted for webs of trust, and similar trust network applications, are largely general and do not cover specifics such as "situational trust" [Marsh,
   1994a], as has been pointed out before [Golbeck et al, 2003]. For instance, agent a_i may
   blindly trust a_j with respect to books, but not trust a_j with respect to trusting others, for a_j
   has been found to accord trust to other people too easily. For our trust propagation scheme
   at hand, we also suppose this largely uni-dimensional concept of trust.

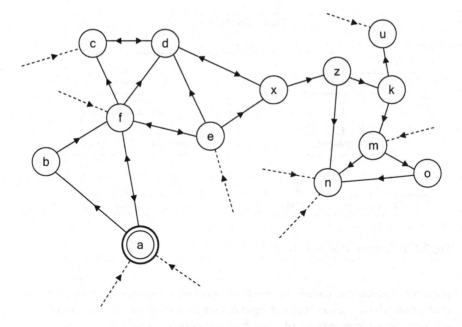

Fig. 7.1 Sample web of trust for agent *a*

7.3 Computational Trust in Social Networks

Trust represents an invaluable and precious good one should award deliberately. Trust metrics compute quantitative *estimates* of how much trust an agent a_i should accord to his peer a_j, taking into account trust ratings from other persons on the network. These metrics should also act "deliberately", not overly awarding trust to persons or agents whose trustworthiness is questionable.

7.3.1 Classification of Trust Metrics

Applications for trust metrics and trust management [Blaze et al, 1996] are rife. First proposals for metrics date back to the early nineties, where trust metrics have been deployed in various projects to support the "Public Key Infrastructure" (PKI) [Zimmermann, 1995]. The metrics proposed by Levien and Aiken [1998], Reiter and Stubblebine [1997b], Maurer [1996], and Beth et al [1994] count among the most popular ones for public key authentication. New areas and research fields apart from PKI have come to make trust metrics gain momentum. Peer-to-peer networks, ubiquitous and mobile computing, and rating systems for online communities, where maintenance of explicit certification authorities is not feasible anymore, have raised the research interest in trust. The whole plethora of available metrics can hereby be defined and characterized along various classification axes. We identify three

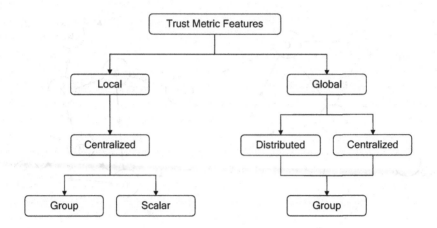

Fig. 7.2 Trust metric classification

principal dimensions, namely *network perspective, computation locus,* and *link evaluation.* These axes are *not* orthogonal, though, for various features impose restrictions on the feature range of other dimensions (see Figure 7.2).

Network Perspective

The first dimension impacts the *semantics* assigned to the values computed. Trust metrics may basically be subdivided into those with *global,* and those with *local* scope. Global trust metrics take into account *all* peers and trust links connecting them. Global trust ranks are assigned to an individual based upon complete trust graph information. Many global trust metrics, such as those presented by Kamvar et al [2003], Guha [2003], and Richardson et al [2003], borrow their ideas from the renowned PageRank algorithm [Page et al, 1998] to compute Web page reputation, and to some lesser extent from HITS [Kleinberg, 1999]. The basic intuition behind these approaches is that nodes should be ranked higher the better the rank of nodes pointing to them. Obviously, the latter notion likewise works for trust and page reputation.

Trust metrics with local scope, on the other hand, take into account personal bias. Interestingly, some researchers claim that only local trust metrics are "true" trust metrics, since global ones compute overall reputation rather than personalized trust[2] [Mui et al, 2002]. Local trust metrics take the agent for whom to compute trust as an additional input parameter and are able to operate on *partial* trust graph information. The rationale behind local trust metrics is that persons an agent a_i trusts may be completely different from the range of individuals that agent a_j deems trustworthy. Local trust metrics exploit structural information defined by personalized

---

[2] Recall the definition of trust given before, expressing that trust is a "subjective expectation".

webs of trust. Hereby, the personal web of trust for individual a_i is given through the set of trust relationships emanating from a_i and passing through nodes he trusts either directly or indirectly, as well as the set of nodes reachable through these relationships. Merging all webs of trust engenders the global trust graph. Local trust metrics comprise Levien's Advogato trust metric [Levien and Aiken, 2000], metrics for modelling the PKI [Beth et al, 1994; Maurer, 1996; Reiter and Stubblebine, 1997b] and the Semantic Web trust infrastructure [Golbeck and Hendler, 2004], and Sun Microsystems's Poblano [Chen and Yeager, 2003]. The latter work borrows from Abdul-Rahman and Hailes [1997].

Computation Locus

The second axis refers to the place where trust relationships between individuals are evaluated and quantified. Local[3] or *centralized* approaches perform all computations in one single machine and hence need to be granted full access to all relevant trust information. The trust data itself may be distributed over the network. Most of the before-mentioned metrics count among the class of centralized approaches.

Distributed metrics for the computation of trust and reputation, such as those described by Richardson et al [2003], Kamvar et al [2003], and Sankaralingam et al [2003], equally deploy the load of computation on every trust node in the network. Upon receiving trust information from his predecessor nodes in the trust graph, an agent a_i merges the data with his own trust assertions and propagates synthesized values to his successor nodes. The entire process of trust computation is necessarily asynchronous and its convergence depends on the eagerness or laziness of nodes to propagate information. Another characteristic feature of distributed trust metrics refers to the fact that they are inherently global. Though the individual computation load is lower with respect to centralized computation approaches, nodes need to store trust information about *any other* node in the system.

Link Evaluation

The third dimension distinguishes scalar and group trust metrics. According to Levien [2004], *scalar* metrics analyze trust assertions independently, while *group trust* metrics evaluate groups of assertions "in tandem". PageRank [Page et al, 1998] and related approaches count among global group trust metrics, for the reputation of one page depends on the ranks of referring pages, thus implying the parallel evaluation of relevant nodes, thanks to mutual dependencies. Advogato [Levien and Aiken, 2000] represents an example for local group trust metrics. Most other trust metrics are scalar ones, tracking trust paths from sources to targets and not performing parallel evaluations of groups of trust assertions. Hence, another basic difference between scalar and group trust metrics refers to their functional design. In general, scalar metrics compute trust between two given individuals a_i and a_j, taken from set A of all agents.

---

[3] Note that in this context, the term "local" refers to the *place of computation* and not the network perspective.

On the other hand, group trust metrics generally compute trust ranks for *sets* of individuals in A. Hereby, global group trust metrics assign trust ranks for every $a_i \in A$, while local ones may also return *ranked subsets* of A. Note that complete trust graph information is only important for *global* group trust metrics, but not for *local* ones. Informally, local group trust metrics may be defined as metrics to compute *neighborhoods* of trusted peers for an individual a_i. As input parameters, these trust metrics take an individual $a_i \in A$ for which to compute the set of peers he should trust, as well as an amount of trust the latter wants to share among his most trustworthy agents. For instance, in [Levien and Aiken, 2000], the amount of trust is said to correspond to the number of agents that a_i wants to trust. The output is hence given by a *trusted subset* of A.

Note that scalar trust metrics are inherently local, while group trust metrics do not impose any restrictions on features for other axes.

7.3.2 Trust and Decentralization

Section 1.1.2 has mentioned the Semantic Web as sample scenario for our decentralized recommender framework. Hence, for the conception of our trust metric, we will also assume the Semantic Web as working environment and representative for large-scale decentralized infrastructures. Note that all considerations presented are also of utter relevance for large, decentralized networks other than the Semantic Web, e.g., very large peer-to-peer networks, the Grid, etc.

Before discussing specific requirements and fitness properties of trust metrics along those axes proposed before, we need to define one common trust *model* on which to rely upon. Some steps towards one such standardized model have already been taken and incorporated into the FOAF [Dumbill, 2002] project. FOAF is an abbreviation for "Friend of a Friend" and aims at enriching personal homepages with machine-readable content encoded in RDF statements. Besides various other information, these publicly accessible pages allow their owners to nominate all individuals part of the FOAF universe they know, thus weaving a "web of acquaintances" [Golbeck et al, 2003]. Golbeck et al [2003] have extended the FOAF schema to also contain *trust* assertions with values ranging from 1 to 9, where 1 denotes complete distrust and 9 absolute trust towards the individual for whom the assertion has been issued. Their assumption that trust and distrust represent *symmetrically opposed* concepts is in line with Abdul-Rahman and Hailes [2000].

The model that we adopt is quite similar to FOAF and its extensions, but only captures the notion of trust and lack of trust, instead of trust and distrust. Note that zero trust and distrust are *not* the same [Marsh, 1994b] and may hence not be intermingled. Explicit modelling of distrust has some serious implications for trust metrics and will hence be discussed separately in Section 7.5. Mind that only few research endeavors have investigated the implementation of distrust so far, e.g., Jøsang et al [2003], Guha [2003], and Guha et al [2004].

7.3.2.1 Trust Model

As is the case for FOAF, we assume that all trust information is publicly accessible for any agent in the system through machine-readable personal homepages distributed over the network. Agents $a_i \in A = \{a_1, a_2, \ldots, a_n\}$ are associated with a partial trust function $W_i \in T = \{W_1, W_2, \ldots, W_n\}$ each, where $W_i : A \to [0,1]^\perp$ holds, which corresponds to the set of trust assertions that a_i has stated.

In most cases, functions $W_i(a_j)$ will be very sparse as the number of individuals an agent is able to assign explicit trust ratings for is much smaller than the total number n of agents. Moreover, the higher the value of $W_i(a_j)$, the more trustworthy a_i deems a_j. Conversely, $W_i(a_j) = 0$ means that a_i considers a_j to be *not trustworthy*. The assignment of trust through continuous values between 0 and 1, and their adopted semantics, is in perfect accordance with [Marsh, 1994a], where possible stratifications of trust values are proposed. Our trust model defines one directed trust graph with nodes being represented by agents $a_i \in A$, and directed edges from nodes a_i to nodes a_j representing trust statements $W_i(a_j)$.

For convenience, we introduce the partial function $W : A \times A \to [0,1]^\perp$, which we define as the union of all partial functions $W_i \in T$.

7.3.2.2 Trust Metrics for Decentralized Networks

Trust and reputation ranking metrics have primarily been used for the PKI [Reiter and Stubblebine, 1997a,b; Levien and Aiken, 1998; Maurer, 1996; Beth et al, 1994], rating and reputation systems part of online communities [Guha, 2003; Levien and Aiken, 2000; Levien, 2004], peer-to-peer networks [Kamvar et al, 2003; Sankaralingam et al, 2003; Kinateder and Rothermel, 2003; Kinateder and Pearson, 2003; Aberer and Despotovic, 2001], and also mobile computing [Eschenauer et al, 2002]. Each of these scenarios favors different trust metrics. For instance, reputation systems for online communities tend to make use of *centralized trust servers* that compute global trust values for all users on the system [Guha, 2003]. On the other hand, peer-to-peer networks of moderate size rely upon distributed approaches that are in most cases based upon PageRank [Kamvar et al, 2003; Sankaralingam et al, 2003].

The Semantic Web, however, as an example for a large-scale decentralized environment, is expected to be made up of millions of nodes, i.e., millions of agents. The fitness of *distributed* approaches to trust metric computation, such as depicted by Richardson et al [2003] and Kamvar et al [2003], hence becomes limited for various reasons:

Trust data storage. Every agent a_i needs to store trust rating information about any other agent a_j on the Semantic Web. Agent a_i uses this information in order to merge it with own trust beliefs and propagates the synthesized information to his trusted agents [Levien, 2004]. Even though one might expect the size of the Semantic Web to be several orders of magnitude smaller than the traditional Web, the number of agents for whom to keep trust information will still exceed the storage capacities of most nodes.

Convergence. The structure of the Semantic Web is diffuse and not subject to some higher ordering principle or hierarchy. Furthermore, the process of trust propagation is *necessarily asynchronous* as there is no central node of authority. As the Semantic Web is huge in size with possibly numerous antagonist or idle agents, convergence of trust values might take a very long time.

The huge advantage of distributed approaches, on the other hand, is the *immediate availability* of computed trust information about any other agent a_j in the system. Moreover, agents have to disclose their trust assertions only to peers they actually *trust* [Richardson et al, 2003]. For instance, suppose that a_i declares his trust in a_j by $W_i(a_j) = 0.1$, which is very low. Hence, a_i might want a_j not to know about that fact. As distributed metrics only propagate *synthesized* trust values from nodes to successor nodes in the trust graph, a_i would not have to openly disclose his trust statements to a_j.

As it comes to centralized, i.e., *locally computed*, metrics, *full* trust information access is required for agents inferring trust. Hence, online communities based on trust require their users to disclose all trust information to the community server, but not necessarily to other peers [Guha, 2003]. Privacy thus remains preserved. On the Semantic Web and in the area of ubiquitous and mobile computing, however, there is no such central authority that computes trust. *Any* agent might want to do so. Our own trust model, as well as trust models proposed by Golbeck et al [2003], Eschenauer et al [2002], and Abdul-Rahman and Hailes [1997], are hence based upon the assumption of *publicly available trust information*. Though privacy concerns may persist, this assumption is vital, owing to the afore-mentioned deficiencies of distributed computation models. Moreover, centralized *global* metrics, such as depicted by Guha [2003] and Page et al [1998], also fail to fit our requirements: because of the huge number of agents issuing trust statements, only dedicated server clusters could be able to manage the whole bulk of trust relationships. For small agents and applications roaming the Semantic Web, global trust computation is not feasible.

Scalar metrics, e.g., PKI proposals [Reiter and Stubblebine, 1997a,b; Levien and Aiken, 1998; Maurer, 1996; Beth et al, 1994] and those metrics described by Golbeck et al [2003], have poor scalability properties, owing to exponential time complexity [Reiter and Stubblebine, 1997a].

Consequently, we advocate *local group* trust metrics for the Semantic Web and other large-scale decentralized networks. These metrics bear several welcome properties with respect to computability and complexity, which may be summarized as follows:

Partial trust graph exploration. Global metrics require a priori full knowledge of the entire trust network. Distributed metrics store trust values for all agents in the system, thus implying massive data storage demands. On the other hand, when computing trusted *neighborhoods*, the trust network only needs to be explored partially: originating from the trust source, one only follows those trust edges that seem promising, i.e., bearing high trust weights, and which are not too far away from the trust source. Inspection of personal, machine-readable homepages is thus performed in a just-in-time fashion. Hence, prefetching bulk trust information is not required.

Computational scalability. Tightly intertwined with partial trust graph exploration is computational complexity. Local group trust metrics scale well to any social network size, as only tiny subsets of relatively constant size[4] are visited. This is not the case for global trust metrics.

7.4 Local Group Trust Metrics

Local group trust metrics, in their function as means to compute trust neighborhoods, have not been subject to mainstream research so far. Significant research has effectively been limited to the work done by Levien [2004] who has conceived and deployed the Advogato group trust metric. This section provides an overview of Advogato and introduces our own Appleseed trust metric, eventually comparing both approaches.

7.4.1 Outline of Advogato Maxflow

The Advogato maximum flow trust metric has been proposed by Levien and Aiken [2000] in order to discover which users are trusted by members of an online community and which are not. Trust is computed through one centralized community server and considered relative to a seed of users enjoying supreme trust. However, the metric is not only applicable to community servers, but also to *arbitrary* agents which may compute *personalized lists* of trusted peers, not only one single global ranking for the whole community they belong to. In this case, the active agent himself constitutes the singleton trust seed. The following paragraphs briefly introduce Advogato's basic concepts. For more detailed information, refer to [Levien and Aiken, 2000], [Levien and Aiken, 1998], and [Levien, 2004].

7.4.1.1 Trust Computation Steps

Local group trust metrics compute sets of agents trusted by those being part of the trust seed. In case of Advogato, its input is given by an integer number n, which is supposed to be equal to the number of members to trust [Levien and Aiken, 2000], as well as the trust seed s, which is a subset of the entire set of users A. The output is a *characteristic function* that maps each member to a boolean value indicating his trustworthiness:

$$\text{Trust}_M : 2^A \times \mathbb{N}_0^+ \rightarrow (A \rightarrow \{\text{true}, \text{false}\}) \tag{7.1}$$

---

[4] Supposing identical parameterizations for the metrics in use, as well as similar network structures.

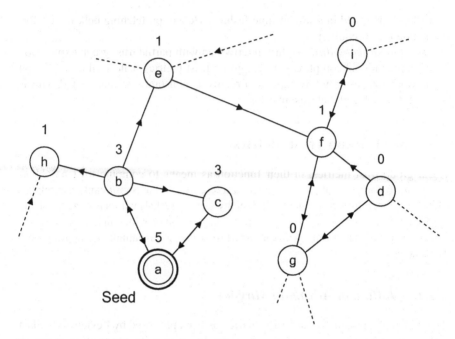

Fig. 7.3 Trust graph *before* conversion for Advogato

The trust model underlying Advogato does *not* provide support for weighted trust relationships in its original version.[5] Hence, trust edges extending from individual x to y express *blind*, i.e., *full*, trust of x in y. The metrics for PKI maintenance suppose similar models. Maximum integer network flow computation [Ford and Fulkerson, 1962] has been investigated by Reiter and Stubblebine [1997b,a] in order to make trust metrics more reliable. Levien adopted and extended this approach for group trust in his Advogato metric:

Capacities $C_A : A \rightarrow \mathbb{N}$ are assigned to every community member $x \in A$ based upon the shortest-path distance from the seed to x. Hereby, the capacity of the seed itself is given by the input parameter n mentioned before, whereas the capacity of each successive distance level is equal to the capacity of the previous level l divided by the average outdegree of trust edges $e \in E$ extending from l. The trust graph we obtain hence contains one single source, which is the set of seed nodes considered as one single "virtual" node, and multiple sinks, i.e., all nodes other than those defining the seed. Capacities $C_A(x)$ constrain nodes. In order to apply Ford-Fulkerson maximum integer network flow [Ford and Fulkerson, 1962], the underlying problem has to be formulated as single-source/single-sink, having capacities $C_E : E \rightarrow \mathbb{N}$ constrain *edges* instead of *nodes*. Hence, Algorithm 7.1 is applied to the old directed graph $G = (A, E, C_A)$, resulting in a new graph structure $G' = (A', E', C_{E'})$.

---

[5] Though various levels of peer certification exist, their interpretation does not perfectly align with weighted trust relationships.

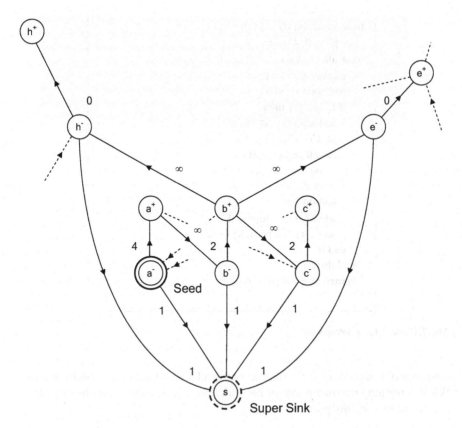

Fig. 7.4 Trust graph *after* conversion for Advogato

Figure 7.4 depicts the outcome of converting node-constrained single-source/multiple-sink graphs (see Figure 7.3) into single-source/single-sink ones with capacities constraining edges.

Conversion is followed by simple integer maximum network flow computation from the trust seed to the super-sink. Eventually, the trusted agents x are exactly those peers for whom there is flow from "negative" nodes x^- to the super-sink. An additional constraint needs to be introduced, requiring flow from x^- to the super-sink whenever there is flow from x^- to x^+. The latter constraint assures that node x does not only serve as an intermediate for the flow to pass through, but is *actually added* to the list of trusted agents when reached by network flow. However, the standard implementation of Ford-Fulkerson traces shortest paths to the sink first [Ford and Fulkerson, 1962]. The above constraint is thus satisfied implicitly already.

Example 4 (Advogato trust computation). Suppose the trust graph depicted in Figure 7.3. The only seed node is a with initial capacity $C_A(a) = 5$. Hence, taking into account the outdegree of a, nodes at unit distance from the seed, i.e., nodes b and c, are assigned capacities $C_A(b) = 3$ and $C_A(c) = 3$, respectively. The average

```
func transform (G = (A,E,C_A)) {
    set E' ← ∅, A' ← ∅;
    for all x ∈ A do
        add node x⁺ to A';
        add node x⁻ to A';
        if C_A(x) ≥ 1 then
            add edge (x⁻,x⁺) to E';
            set C_{E'}(x⁻,x⁺) ← C_A(x) − 1;
            for all (x,y) ∈ E do
                add edge (x⁺,y⁻) to E';
                set C_{E'}(x⁺,y⁻) ← ∞;
            end do
            add edge (x⁻, supersink) to E';
            set C_{E'}(x⁻, supersink) ← 1;
        end if
    end do
    return G' = (A',E',C_{E'});
}
```

Alg. 7.1 Trust graph conversion

outdegree of both nodes is 2.5 so that second level nodes e and h obtain unit capacity. When computing maximum integer network flow, agent a will accept himself, b, c, e, and h as trustworthy peers.

7.4.1.2 Attack-Resistance Properties

Advogato has been designed with resistance against massive attacks from malicious agents outside of the community in mind. Therefore, an upper bound for the number of "bad" peers chosen by the metric is provided in [Levien and Aiken, 2000], along with an informal security proof to underpin its fitness. Resistance against malevolent users trying to break into the community can already be observed in the example depicted by Figure 7.1, supposing node n to be "bad": though agent n is trusted by numerous persons, he is deemed less trustworthy than, for instance, x. While there are fewer agents trusting x, these agents enjoy higher trust reputation[6] than the numerous persons trusting n. Hence, it is not just the *number* of agents trusting an individual i, but also the trust *reputation* of these agents that exerts an impact on the trust assigned to i. PageRank [Page et al, 1998] works in a similar fashion and has been claimed to possess properties of attack-resistance similar to those of the Advogato trust metric [Levien, 2004]. In order to make the concept of attack-resistance more tangible, Levien proposes the "bottleneck property" as a common feature of attack-resistant trust metrics. Informally, this property states that the "trust

---

[6] With respect to seed node a.

quantity accorded to an edge $s \to t$ is not significantly affected by changes to the successors of t" [Levien, 2004].

Attack-resistance features of various trust metrics are discussed in detail in [Levien and Aiken, 1998] and [Twigg and Dimmock, 2003].

7.4.2 The Appleseed Trust Metric

The Appleseed trust metric constitutes the main contribution of this chapter and is our novel proposal for local group trust metrics. In contrast to Advogato, being inspired by maximum network flow computation, the basic intuition of Appleseed is motivated by *spreading activation models*. Spreading activation models have first been proposed by Quillian [1968] in order to simulate human comprehension through semantic memory, and are commonly described as "models of retrieval from long-term memory in which activation subdivides among paths emanating from an activated mental representation" [Smith et al, 2003]. By the time of this writing, the seminal work of Quillian has been ported to a whole plethora of other disciplines, such as latent semantic indexing [Ceglowski et al, 2003] and text illustration [Hartmann and Strothotte, 2002]. As an example, we will briefly introduce the spreading activation approach adopted by Ceglowski et al [2003], used for semantic search in contextual network graphs, in order to then relate Appleseed to that work.

7.4.2.1 Searches in Contextual Network Graphs

The graph model underlying contextual network search graphs is almost identical in structure to the one presented in Section 7.3.2.1, i.e., edges $(x, y) \in E \subseteq A \times A$ connecting nodes $x, y \in A$. Edges are assigned continuous weights through $W : E \to [0, 1]$. Source node s, the node from which we start searching, is activated through an injection of energy e, which is then propagated to other nodes along edges according to some set of simple rules: all energy is *fully divided* among successor nodes with respect to their normalized local edge weight, i.e., the higher the weight of an edge $(x, y) \in E$, the higher the portion of energy that flows along that edge. Furthermore, supposing average outdegrees greater than one, the closer node x to the injection source s, and the more paths lead from s to x, the higher the amount of energy flowing into x. To eliminate endless, marginal and negligible flow, energy streaming into node x must exceed threshold T in order not to run dry. The described approach is captured formally by Algorithm 7.2, which propagates energy recursively.

7.4.2.2 Trust Propagation

Algorithm 7.2 shows the basic intuition behind spreading activation models. In order to tailor these models to trust computation, later to become the Appleseed trust metric, serious adaptations are necessary. For instance, procedure energize(e, s) registers *all* energy e that has passed through node x, stored in energy(x). Hence, energy(x) represents the *relevance rank* of x. Higher values indicate higher node rank. However, at the same time, all energy contributing to the rank of x is passed

```
procedure energize (e ∈ ℝ₀⁺, s ∈ A) {
    energy(s) ← energy(s) + e;
    e' ← e / Σ₍s,n₎∈E W(s,n);
    if e > T then
        ∀(s,n) ∈ E : energize (e' · W(s,n),n);
    end if
}
```

Alg. 7.2 Recursive energy propagation

without loss to successor nodes. Interpreting energy ranks as trust ranks thus implies numerous issues of semantic consistency as well as computability. Consider the graph depicted in Figure 7.5(a). Applying spreading activation according to Ceglowski et al [2003], trust ranks of nodes b and d will be identical. However, intuitively, d should be accorded *less* trust than b, since d's shortest-path distance to the trust seed is higher. Trust decay is commonly agreed upon [Guha, 2003; Jøsang et al, 2003], for people tend to trust individuals trusted by immediate friends more than individuals trusted only by friends of friends. Figure 7.5(b) depicts even more serious issues: all energy, or trust[7], respectively, distributed along edge (a,b) becomes *trapped in a cycle* and will never be accorded to any other nodes but those being part of that cycle, i.e., b, c, and d. These nodes will eventually acquire infinite trust rank. Obviously, the *bottleneck property* [Levien, 2004] does not hold. Similar issues occur with simplified versions of PageRank [Page et al, 1998], where cycles accumulating infinite rank have been dubbed "rank sinks".

7.4.2.3 Spreading Factor

We handle both issues, i.e., trust decay in node chains and elimination of rank sinks, by tailoring the algorithm to rely upon our global *spreading factor* d. Hereby, let $in(x)$ denote the energy influx into node x. Parameter d then denotes the portion of energy $d \cdot in(x)$ that node x distributes among successors, while retaining $(1 - d) \cdot in(x)$. For instance, suppose $d = 0.85$ and energy quantity $in(x) = 5.0$ flowing into node x. Then, the total energy distributed to successor nodes amounts to 4.25, while the energy rank energy(x) of x increases by 0.75. Special treatment is necessary for nodes with zero outdegree. For simplicity, we assume all nodes to have an outdegree of at least one, which makes perfect sense, as will be shown later.

The spreading factor concept is very intuitive and, in fact, very close to real models of energy spreading through networks. Observe that the overall amount of energy in the network, after initial activation in^0, does not change over time. More formally, suppose that energy$(x) = 0$ for all $x \in A$ before injection in^0 into source s. Then the

---

[7] The terms "energy" and "trust" are used interchangeably in this context.

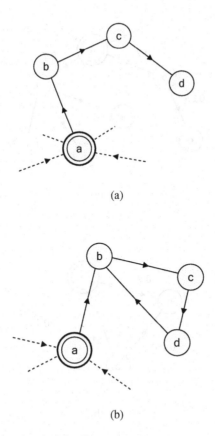

(a)

(b)

Fig. 7.5 Node chains (a) and rank sinks (b)

following equation holds in every computation step of our modified spreading algorithm, incorporating the concept of spreading factor d:

$$\sum_{x \in A} \text{energy}(x) = \text{in}^0 \qquad (7.2)$$

Spreading factor d may also be seen as the *ratio* between *direct* trust in x and trust in the ability of x to *recommend* others as trustworthy peers. For instance, Beth et al [1994] and Maurer [1996] explicitly differentiate between *direct* trust edges and *recommendation* edges.

We commonly assume $d = 0.85$, though other values may also seem reasonable. For instance, having $d \leq 0.5$ allows agents to keep most of the trust they are granted for themselves and only pass small portions of trust to their peers. Observe that low values for d favor trust proximity to the source of trust injection, while high values allow trust to also reach more distant nodes. Furthermore, the introduction of spreading factor d is crucial for making Appleseed retain Levien's bottleneck property, as will be shown in later sections.

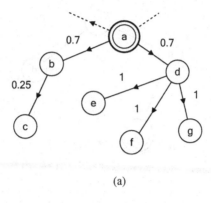

(a)

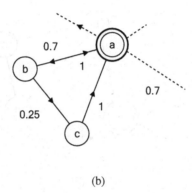

(b)

Fig. 7.6 Issues with trust normalization

7.4.2.4   Rank Normalization

Algorithm 7.2 makes use of edge weight normalization, i.e., the quantity $e_{x \to y}$ of energy distributed along (x, y) from x to successor node y depends on the *relative* weight of $x \to y$, i.e., $W(x, y)$ compared to the sum of weights of all outgoing edges of x:

$$e_{x \to y} = d \cdot \text{in}(x) \cdot \frac{W(x, y)}{\sum_{(x,s) \in E} W(x, s)} \qquad (7.3)$$

Normalization is common practice in many trust metrics, among those PageRank [Page et al, 1998], EigenTrust [Kamvar et al, 2003], and AORank [Guha, 2003]. However, while normalized reputation or trust seem reasonable for models with plain, non-weighted edges, serious interferences occur when edges are *weighted*, as is the case for our trust model adopted in Section 7.3.2.1.

For instance, refer to Figure 7.6(a) for unwanted effects: The amounts of energy that node a accords to successors b and d, i.e., $e_{a \to b}$ and $e_{a \to d}$, respectively, are identical in value. Note that b has issued only *one* trust statement $W(b,c) = 0.25$, stating that b's trust in c is rather weak. On the other hand, d assigns *full* trust to individuals e, f, and g. Nevertheless, the overall trust rank for d will be much higher than for any successor of d, for c is accorded $e_{a \to b} \cdot d$, while e, f, and g only obtain $e_{a \to d} \cdot d \cdot 1/3$ each. Hence, c will be trusted *three times* as much as e, f, and g, which is not reasonable at all.

7.4.2.5 Backward Trust Propagation

The above issue has already been discussed by Kamvar et al [2003], but no solution has been proposed therein, arguing that "substantially good results" have been achieved despite the drawbacks. We propose to alleviate the problem by making use of *backward propagation* of trust to the source: when metric computation takes place, additional "virtual" edges (x,s) from every node $x \in A \setminus \{s\}$ to the trust source s are created. These edges are assigned full trust $W(x,s) = 1$. Existing backward links (x,s), along with their weights, are "overwritten". Intuitively, every node is supposed to *blindly trust the trust source* s, see Figure 7.6(b). The impacts of adding backward propagation links are threefold:

Mitigating relative trust.   Again, we refer to Figure 7.6(a). Trust distribution in the underlying case becomes much fairer through backward propagation links, for c now only obtains $e_{a \to b} \cdot d \cdot (0.25/(1+0.25))$ from source s, while e, f, and g are accorded $e_{a \to d} \cdot d \cdot (1/4)$ each. Hence, trust ranks of both e, f, and g amount to 1.25 times the trust assigned to c.

Avoidance of dead ends.   Dead ends, i.e., nodes x with zero outdegree, require special treatment in our computation scheme. Two distinct approaches may be adopted. First, the portion of incoming trust $d \cdot \text{in}(x)$ supposed to be passed to successor nodes is completely discarded, which contradicts our intuition of no energy leaving the system. Second, instead of retaining $(1 - d) \cdot \text{in}(x)$ of incoming trust, x keeps *all* trust. The latter approach is also not sensible as it encourages users to not issue trust statements for their peers. Luckily, with backward propagation of trust, all nodes are *implicitly linked* to the trust source s, so that there are no more dead ends to consider.

Favoring trust proximity.   Backward links to the trust source s are favorable for nodes close to the source, as their eventual trust rank will increase. On the other hand, nodes further away from s are penalized.

7.4.2.6 Nonlinear Trust Normalization

In addition to backward propagation, we propose supplementary measures to decrease the negative impact of trust spreading based on relative weights. Situations where nodes y with poor ratings from x are awarded high overall trust ranks, thanks to the low outdegree of x, have to be avoided. Taking the squares of local trust weights provides an appropriate solution:

$$e_{x \to y} = d \cdot \text{in}(x) \cdot \frac{W(x,y)^2}{\sum_{(x,s) \in E} W(x,s)^2} \qquad (7.4)$$

As an example, refer to node b in Figure 7.6(b). With squared normalization, the total amount of energy flowing backward to source a increases, while the amount of energy flowing to the poorly trusted node c decreases significantly. Accorded trust quantities $e_{b \to a}$ and $e_{b \to c}$ amount to $d \cdot \text{in}(b) \cdot (1/1.0625)$ and $d \cdot \text{in}(b) \cdot (0.0625/1.0625)$, respectively. A more severe penalization of poor trust ratings can be achieved by selecting powers above two.

7.4.2.7 Algorithm Outline

Having identified modifications to apply to spreading activation models in order to tailor them for local group trust metrics, we are now able to formulate the core algorithm of Appleseed. Input and output are characterized as follows:

$$\text{Trust}_\alpha : A \times \mathbb{R}_0^+ \times [0,1] \times \mathbb{R}^+ \to (\text{trust} : A \to \mathbb{R}_0^+) \qquad (7.5)$$

The first input parameter specifies trust seed s, the second trust injection e, parameter three identifies spreading factor $d \in [0,1]$, and the fourth argument binds accuracy threshold T_c, which serves as convergence criterion. Similar to Advogato, the output is an assignment function of trust with domain A. However, Appleseed allows *rankings* of agents with respect to trust accorded. Advogato, on the other hand, only assigns boolean values indicating presence or absence of trust.

Appleseed works with *partial* trust graph information. Nodes are accessed only when needed, i.e., when reached by energy flow. Trust ranks trust(x), which correspond to energy(x) in Algorithm 7.2, are initialized to 0. Any unknown node u hence obtains trust$(u) = 0$. Likewise, virtual trust edges for backward propagation from node x to the source are added *at the moment that x is discovered*. In every iteration, for those nodes x reached by flow, the amount of incoming trust is computed as follows:

$$\text{in}(x) = d \cdot \sum_{(p,x) \in E} \left(\text{in}(p) \cdot \frac{W(p,x)}{\sum_{(p,s) \in E} W(p,s)} \right) \qquad (7.6)$$

Incoming flow for x is hence determined by all flow that predecessors p distribute along edges (p,x). Note that the above equation makes use of *linear normalization* of relative trust weights. The replacement of linear by nonlinear normalization according to Section 7.4.2.6 is straight-forward, though. The trust rank of x is updated as follows:

$$\text{trust}(x) \leftarrow \text{trust}(x) + (1 - d) \cdot \text{in}(x) \qquad (7.7)$$

Trust networks generally contain cycles and thus allow no topological sorting of nodes. Hence, the computation of $in(x)$ for reachable $x \in A$ becomes *inherently recursive*. Several iterations for all nodes are required in order to make the computed information converge towards the least fixpoint. The following criterion has to be satisfied for convergence, relying upon accuracy threshold T_c briefly introduced before.

Definition 1 (Termination). Suppose that $A_i \subseteq A$ represents the set of nodes that were discovered until step i, and $\mathrm{trust}_i(x)$ the current trust ranks for all $x \in A$. Then the algorithm terminates when the following condition is satisfied after step i:

$$\forall x \in A_i : \mathrm{trust}_i(x) - \mathrm{trust}_{i-1}(x) \leq T_c \tag{7.8}$$

Informally, Appleseed terminates when changes of trust ranks with respect to the preceding iteration $i - 1$ are not greater than accuracy threshold T_c.

Moreover, when supposing spreading factor $d > 0$, accuracy threshold $T_c > 0$, and trust source s part of some connected component $G' \subseteq G$ containing at least two nodes, convergence, and thus termination, is guaranteed. The following paragraph gives an informal proof:

Proof (Convergence of Appleseed). Assume that f_i denotes step i's quantity of energy flowing through the network, i.e., all the trust that has not been captured by some node x through function $\mathrm{trust}_i(x)$. From Equation 7.2 follows that in^0 constitutes the *upper boundary* of trust energy floating through the network, and f_i can be computed as follows:

$$f_i = in^0 - \sum_{x \in A} \mathrm{trust}_i(x) \tag{7.9}$$

Since $d > 0$ and $\exists (s,x) \in E, x \neq s$, the sum of the current trust ranks $\mathrm{trust}_i(x)$ of all $x \in A$ is *strictly increasing* for increasing i. Consequently, $\lim_{i \to \infty} f_i = 0$ holds. Moreover, since termination is defined by some fixed accuracy threshold $T_c > 0$, there exists some step k such that $\lim_{i \to k} f_i \leq T_c$.

7.4.2.8 Parameterization and Experiments

Appleseed allows numerous parameterizations of input variables, some of which are subject to discussion in the section at hand. Moreover, we provide experimental results exposing the observed effects of parameter tuning. Note that all experiments have been conducted on data obtained from "real" social networks: we have written several Web crawling tools to mine the Advogato community Web site and extract trust assertions stated by its more than $8,000$ members. Hereafter, we converted all trust data to our trust model proposed in Section 7.3.2.1. The Advogato community server supports four different levels of peer certification, namely OBSERVER, APPRENTICE, JOURNEYER, and MASTER. We mapped these *qualitative* certification levels to quantitative ones, assigning $W(x,y) = 0.25$ for x certifying y as OBSERVER,

```
func Trustα (s ∈ A, in⁰ ∈ ℝ₀⁺, d ∈ [0,1], Tc ∈ ℝ⁺) {
    set in₀(s) ← in⁰, trust₀(s) ← 0, i ← 0;
    set A₀ ← {s};
    repeat
        set i ← i + 1;
        set Aᵢ ← Aᵢ₋₁;
        ∀x ∈ Aᵢ₋₁ : set inᵢ(x) ← 0;
        for all x ∈ Aᵢ₋₁ do
            set trustᵢ(x) ← trustᵢ₋₁(x) + (1 − d) · inᵢ₋₁(x);
            for all (x,u) ∈ E do
                if u ∉ Aᵢ then
                    set Aᵢ ← Aᵢ ∪ {u};
                    set trustᵢ(u) ← 0, inᵢ(u) ← 0;
                    add edge (u,s), set W(u,s) ← 1;
                end if
                set w ← W(x,u) / Σ₍ₓ,ᵤ'₎∈E W(x,u');
                set inᵢ(u) ← inᵢ(u) + d · inᵢ₋₁(x) · w;
            end do
        end do
        set m = maxᵧ∈Aᵢ {trustᵢ(y) − trustᵢ₋₁(y)};
    until (m ≤ Tc)
    return (trust : {(x, trustᵢ(x)) | x ∈ Aᵢ});
}
```

Alg. 7.3 Outline of the Appleseed trust metric

$W(x,y) = 0.5$ for an APPRENTICE, and so forth. The Advogato community undergoes rapid growth and our crawler extracted $3,224,101$ trust assertions. Preprocessing and data cleansing were thus inevitable, eliminating reflexive trust statements $W(x,x)$ and shrinking trust certificates to reasonable sizes. Note that some eager Advogato members have issued *more than two thousand* trust statements, yielding an overall average outdegree of 397.69 assertions per node.Clearly, this figure is beyond dispute. Hence, applying our set of extraction tools, we tailored the test data obtained from Advogato to our needs and extracted trust networks with specific average outdegrees for the experimental analysis.

Trust Injection

Trust values $trust(x)$ computed by the Appleseed metric for source s and node x may differ greatly from explicitly assigned trust weights $W(s,x)$. We already mentioned before that computed trust ranks may *not* be interpreted as absolute values, but rather in comparison with ranks assigned to all other peers. In order to make assigned rank values more tangible, though, one might expect that tuning the trust injection in^0

```
func Trustheu (s ∈ A, d ∈ [0, 1], Tc ∈ ℝ+) {
    add node i, edge (s, i), set W(s, i) ← 1;
    set in⁰ ← 20, ε ← 0.1;
    repeat
        set trust ← Trustα (s, in⁰, d, Tc);
        in⁰ ← adapt (W(s, i), trust(i), in⁰);
    until trust(i) ∈ [W(s, i) − ε, W(s, i) + ε]
    remove node i, remove edge (s, i);
    return Trustα (s, in⁰, d, Tc);
}
```

Alg. 7.4 Heuristic weight alignment method

to satisfy the following proposition will align computed ranks and explicit trust
statements:

$$\forall (s, x) \in E: \text{ trust}(x) \in [W(s, x) - \varepsilon, W(s, x) + \varepsilon] \tag{7.10}$$

However, when assuming reasonably small ε, the approach does not succeed. Recall that *computed* trust values of successor nodes x of s do not only depend on
assertions made by s, but also on trust ratings asserted by other peers. Hence, a perfect alignment of explicit trust ratings with computed ones cannot be accomplished.
However, we propose a heuristic alignment method, incorporated into Algorithm
7.4, which has proven to work remarkably well in diverse test scenarios. The basic idea is to add another node i and edge (s, i) with $W(s, i) = 1$ to the trust graph
$G = (A, E, W)$, treating (s, i) as an indicator to test whether trust injection in^0 is
"good" or not. Consequently, parameter in^0 has to be adapted in order to make
trust(i) converge towards $W(s, i)$. The trust metric computation is hence repeated
with different values for in^0 until convergence of the explicit and the computed trust
value of i is achieved. Eventually, edge (s, i) and node i are removed and the computation is performed one more time. Experiments have shown that our imperfect
alignment method yields computed ranks trust(x) for direct successors x of trust
source s which come close to previously specified trust statements $W(s, x)$.

Spreading Factor

Small values for d tend to overly reward nodes close to the trust source and penalize
remote ones. Recall that *low d* allows nodes to retain most of the incoming trust
quantity for themselves, while *large d* stresses the recommendation of trusted individuals and makes nodes distribute most of the assigned trust to their successor
nodes.

Experiment 1 (Spreading factor impact). We compare distributions of computed
rank values for three diverse instantiations of d, namely $d_1 = 0.1$, $d_2 = 0.5$, and
$d_3 = 0.85$. Our setup is based upon a social network with an average outdegree

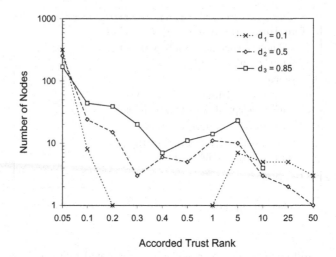

Fig. 7.7 Spreading factor impact

of 6 trust assignments, and features 384 nodes reached by trust energy spreading from our designated trust source. We furthermore suppose $in^0 = 200$, $T_c = 0.01$, and *linear* weight normalization. Computed ranks are classified into 11 histogram cells with nonlinear cell width. Obtained output results are displayed in Figure 7.7. Mind that we have chosen *logarithmic* scales for the vertical axis in order to render the diagram more legible. For d_1, we observe that the largest number of nodes x with ranks $trust(x) \geq 25$ is generated. On the other hand, virtually no ranks ranging from 0.2 to 1 are assigned, while the number of nodes with ranks smaller than 0.05 is again much higher for d_1 than for both d_2 and d_3. Instantiation $d_3 = 0.85$ exhibits behavior opposed to that of d_1. No ranks with $trust(x) \geq 25$ are accorded, while interim ranks between 0.1 and 10 are much more likely for d_3 than for both other instantiations of spreading factor d. Consequently, the number of ranks below 0.05 is lowest for d_3.

The experiment demonstrates that high values for parameter d tend to distribute trust more evenly, neither overly rewarding nodes close to the source, nor penalizing remote ones too rigidly. On the other hand, low d assigns high trust ranks to very few nodes, namely those which are closest to the source, while the majority of nodes obtains very low trust rank. We propose to set $d = 0.85$ for general use.

Convergence

We already mentioned before that the Appleseed algorithm is *inherently recursive*. Parameter T_c represents the ultimate criterion for termination. We demonstrate through an experiment that convergence is reached very fast, no matter how large the number of nodes trust is flowing through, and no matter how large the initial trust injection.

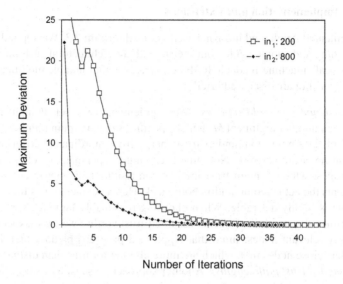

Fig. 7.8 Convergence of Appleseed

Experiment 2 (Convergence rate). The trust network we consider has an average outdegree of 5 trust statements per node. The number of nodes for which trust ranks are assigned amounts to 572. We suppose $d = 0.85$, $T_c = 0.01$, and *linear* weight normalization. Two separate runs were computed, one with trust activation $in_1 = 200$, the other with initial energy $in_2 = 800$. Figure 7.8 demonstrates the rapid convergence of both runs. Though the trust injection for the second run is 4 times as high as for the first, convergence is reached in only few more iterations: run one takes 38 iterations, run two terminates after 45 steps.

For both runs, we assumed accuracy threshold $T_c = 0.01$, which is extremely small and accurate beyond necessity already. However, experience taught us that convergence takes place rapidly even for very large networks and high amounts of trust injected, so that assuming the latter value for T_c poses no scalability issues. In fact, the amount of nodes taken into account for trust rank assignment in the above example well exceeds practical usage scenarios: mind that the case at hand demands 572 files to be fetched from the Web, complaisantly supposing that these pages are cached after their first access. Hence, we claim that the actual bottleneck of group trust computation is *not* the Appleseed metric itself, but downloads of trust resources from the network. This bottleneck might also be the reason for selecting thresholds T_c greater than 0.01, in order to make the algorithm terminate after fewer node accesses.

7.4.2.9   Implementation and Extensions

We implemented Appleseed in Java, based upon Algorithm 7.3. We applied moderate fine-tuning and supplemented our metric with an architectural cushion in order to access "real" machine-readable RDF homepages. Other notable modifications to the core algorithm are discussed briefly:

Maximum number of unfolded nodes. We supplemented the set of input parameters by yet another argument M, which specifies the maximum number of nodes to unfold. This extension hinders trust energy from inordinately covering major parts of the entire network. Note that accessing the personal, machine-readable homepages, which contain trust information required for metric computation, represents the actual computation bottleneck. Hence, expanding as few nodes as possible is highly desirable. When choosing reasonably large M, for instance, three times the number of agents assumed trustworthy, we may expect to not miss any relevant nodes: mind that Appleseed proceeds breadth-first and thus considers close nodes first, which are more eligible for trust than distant ones.

Upper-bounded trust path lengths. Another approach to sensibly restrict the number of nodes unfolded relies upon upper-bounded path lengths. The idea of constraining path lengths for trust computation has been adopted before by Reiter and Stubblebine [1997a] and within the X.509 protocol [Housely et al, 1999]. Depending on the overall trust network connectivity, we opt for maximum path lengths around three, aware of Milgram's "six degrees of separation" paradigm [Milgram, 1992]. In fact, trust decay is inherent to Appleseed, thanks to spreading factor d and backward propagation. Stripping nodes at large distances from the seed therefore only marginally affects the trust metric computation results while simultaneously providing major speed-ups.

Zero trust retention for the source. Third, we modified Appleseed to hinder trust source s from accumulating trust energy, essentially introducing one novel spreading factor $d_s = 1.0$ for the seed only. Consequently, all trust is divided among peers of s and none retained, which is reasonable. Convergence may be reached faster, since $\text{trust}_{i+1}(x) - \text{trust}_i(x)$ tends to be maximal for seed node s, thanks to backward propagation of trust (see Section 7.4.2.5). Furthermore, supposing the same trust quantity in^0 injected, assigned trust ranks become greater in value, also enlarging gaps between neighbors in trust rank.

Testbed Conception

Trust metrics and models for trust propagation have to be *intuitive*, i.e., humans must eventually comprehend *why* agent a_i has been accorded a higher trust rank than a_j and come to similar results when asked for personal judgement. Consequently, we implemented our own testbed, which graphically displays social networks. We made use of the YFILES [Wiese et al, 2001] library to perform complex graph drawing and layouting tasks[8]. The testbed allows for parameterizing Appleseed through dialogs.

---

[8] See Figure 9.2 for a sample visualization.

Detailed output is provided, both graphical and textual. Graphical results comprise the highlighting of nodes with trust ranks above certain thresholds, while textual results return quantitative trust ranks of all accessed nodes, the number of iterations, and so forth. We also implemented the Advogato trust metric and incorporated the latter into our testbed. Hereby, our implementation of Advogato does not require a priori complete trust graph information, but accesses nodes "just in time", similar to Appleseed. All experiments were conducted on top of the testbed application.

7.4.3 Comparison of Advogato and Appleseed

Advogato and Appleseed are both implementations of local group trust metrics. Advogato has already been successfully deployed into the Advogato online community, though quantitative evaluation results have not been provided yet. In order to evaluate the fitness of Appleseed as an appropriate means for group trust computation, we relate our approach to Advogato for qualitative comparison:

(F.1) Attack-resistance. This property defines the behavior of trust metrics in case of malicious nodes trying to invade into the system. For evaluation of attack-resistance capabilities, we have briefly introduced the "bottleneck property" in Section 7.4.1.2, which holds for Advogato. In order to recapitulate, suppose that s and t are nodes and connected through trust edge (s, t). Node s is assumed good, while t is an attacking agent trying to make good nodes trust malevolent ones. In case the bottleneck property holds, manipulation "on the part of bad nodes does not affect the trust value" [Levien, 2004]. Clearly, Appleseed satisfies the bottleneck property, for nodes cannot raise their impact by modifying the structure of trust statements they issue. Bear in mind that the amount of trust accorded to agent t *only* depends on his predecessors and does not increase when t adds more nodes. Both, spreading factor d and normalization of trust statements, ensure that Appleseed maintains attack-resistance properties according to Levien's definition.

(F.2) Eager truster penalization. We have indicated before that issuing multiple trust statements dilutes trust accorded to successors. According to Guha [2003], this does not comply with real world observations, where statements of trust "do not decrease in value when the user trusts one more person [...]". The malady that Appleseed suffers from is common to many trust metrics, most notably those based upon finding principal eigenvectors [Page et al, 1998; Kamvar et al, 2003; Richardson et al, 2003]. On the other hand, the approach pursued by Advogato does *not* penalize trust relationships asserted by eager trust dispensers, for node capacities do not depend on *local* information. Remember that capacities of nodes pertaining to level l are assigned based on the capacity of level $l - 1$, as well as the *overall* outdegree of nodes part of that level. Hence, Advogato *encourages* agents issuing numerous trust statements, while Appleseed *penalizes* overly abundant trust certificates.

(F.3) Deterministic trust computation. Appleseed is deterministic with respect to the assignment of trust rank to agents. Hence, for any arbitrary trust graph

$G = (A, E, W)$ and for every node $x \in A$, linear equations allow for character-
izing the amount of trust assigned to x, as well as the quantity that x accords
to successor nodes. Advogato, however, is *non-deterministic*. Though the *number* of trusted agents, and therefore the computed maximum flow size, is determined for given input parameters, the set of agents is not. Changing the order in
which trust assertions are issued may yield different results. For example, suppose $C_A(s) = 1$ holds for trust seed s. Furthermore, assume s has issued trust
certificates for two agents, b and c. The actual choice between b or c as trustworthy peer with maximum flow *only depends on the order* in which nodes are
accessed.

(F.4) Model and output type. Basically, Advogato supports non-weighted trust
statements only. Appleseed is more versatile by virtue of its trust model based
on *weighted* trust certificates. In addition, Advogato returns one set of trusted
peers, whereas Appleseed assigns *ranks* to agents. These ranks allow to select
most trustworthy agents first and relate them to each other with respect to their
accorded rank. Hereby, the definition of thresholds for trustworthiness is left to
the user who can thus tailor relevant parameters to fit different application scenarios. For instance, raising the application-dependent threshold for the selection of
trustworthy peers, which may be either an absolute or a relative value, allows for
enlarging the neighborhood of trusted peers. Appleseed is hence more adaptive
and flexible than Advogato.

The afore-mentioned characteristics of Advogato and Appleseed are briefly summarized in Table 7.1.

Table 7.1 Characteristics of Advogato and Appleseed

	Feature **F.1**	Feature **F.2**	Feature **F.3**	Feature **F.4**
Advogato	yes	no	no	boolean
Appleseed	yes	yes	yes	ranking

7.5 Distrust

The notion of distrust is one of the most controversial topics when defining trust
metrics and trust propagation. Most approaches completely *ignore* distrust and only
consider *full* trust or *degrees of trust* [Levien and Aiken, 1998; Mui et al, 2002; Beth
et al, 1994; Maurer, 1996; Reiter and Stubblebine, 1997a; Richardson et al, 2003].
Others, among those Abdul-Rahman and Hailes [1997], Chen and Yeager [2003],
Aberer and Despotovic [2001], and Golbeck et al [2003], allow for distrust ratings,
though, but do not consider the subtle semantic differences that exist between those

two notions, i.e., trust and distrust. Consequently, according to Gans et al [2001], "distrust is regarded as just the other side of the coin, that is, there is generally a symmetric scale with complete trust on one end and absolute distrust on the other." Furthermore, some researchers equate the notion of distrust with *lack of trust information*. However, in his seminal work on the essence of trust, Marsh [1994a] has already pointed out that those two concepts, i.e., lack of trust and distrust, may *not* be intermingled. For instance, in absence of trustworthy agents, one might be more prone to accept recommendations from non-trusted persons, being non-trusted probably because of lack of prior experiences [Marsh, 1994b], than from persons we explicitly *distrust*, the distrust resulting from bad past experiences or deceit. However, even Marsh pays little attention to the specifics of distrust.

Gans et al [2001] were among the first to recognize the importance of distrust, stressing the fact that "distrust is an irreducible phenomenon that cannot be offset against any other social mechanisms", including trust. In their work, an explicit distinction between confidence, trust, and distrust is made. Moreover, the authors indicate that distrust might be highly relevant to social networks. Its impact is not inherently negative, but may also influence the network in an extremely positive fashion. However, the primary focus of this work is on methodology issues and planning, not considering trust assertion evaluations and propagation through appropriate metrics.

Guha et al [2004] acknowledge the immense role of distrust with respect to trust propagation applications, arguing that "distrust statements are very useful for users to debug their web of trust" [Guha, 2003]. For example, suppose that agent a_i blindly trusts a_j, which again blindly trusts a_k. However, a_i completely distrusts a_k. The distrust statement hence ensures that a_i will *not* accept beliefs and ratings from a_k, irrespective of him trusting a_j trusting a_k.

7.5.1 Semantics of Distrust

The non-symmetrical nature of distrust and trust, being two dichotomies, has already been recognized by recent sociological research [Lewicki et al, 1998]. In this section, we investigate the differences between distrust and trust with respect to inference opportunities and the propagation of beliefs.

7.5.1.1 Distrust as Negated Trust

Interpreting distrust as the negation of trust has been adopted by many trust metrics, among those trust metrics proposed by Abdul-Rahman and Hailes [1997, 2000], Jøsang et al [2003], and Chen and Yeager [2003]. Basically, these metrics compute trust values by analyzing *chains* of trust statements from source s to target t, eventually merging them to obtain an aggregate value. Each chain hereby becomes synthesized into one single number through *weighted multiplication* of trust values along trust paths. Serious implications resulting from the assumption that trust concatenation relates to multiplication [Richardson et al, 2003], and distrust to negated trust, arise when agent a_i distrusts a_j, who distrusts a_k:[9]

$$\neg\,\text{trust}(a_i, a_j) \wedge \neg\,\text{trust}(a_j, a_k) \models \text{trust}(a_i, a_k) \qquad (7.11)$$

Jøsang et al [2003] are aware of this rather unwanted effect, but do not question its correctness, arguing that "the enemy of your enemy could well be your friend". Guha [2003], on the other hand, indicates that two distrust statements cancelling out each other commonly does *not* reflect desired behavior.

7.5.1.2 Propagation of Distrust

The *conditional transitivity* of trust [Abdul-Rahman and Hailes, 1997] is commonly agreed upon and represents the foundation and principal premiss that trust metrics rely upon. However, no consensus in literature has been achieved with respect to the *degree* of transitivity and the decay rate of trust. Many approaches therefore explicitly distinguish between *recommendation* trust and *direct* trust [Jøsang et al, 2003; Abdul-Rahman and Hailes, 1997; Maurer, 1996; Beth et al, 1994; Chen and Yeager, 2003] in order to keep apart the transitive fraction of trust from the non-transitive. Hence, in these works, only the *ultimate* edge within the trust chain, i.e., the one linking to the trust target, needs to be direct, while all others are supposed to be recommendations. For the Appleseed trust metric, this distinction is made through the introduction of spreading factor d. However, the conditional transitivity property of trust does not equally extend to distrust. The case of double negation through distrust propagation has already been considered. Now suppose, for instance, that a_i distrusts a_j, who trusts a_k. Supposing distrust to propagate through the network, we come to make the following inference:

$$\text{distrust}(a_i, a_j) \wedge \text{trust}(a_j, a_k) \models \text{distrust}(a_i, a_k) \qquad (7.12)$$

The above inference is more than questionable, for a_i penalizes a_k simply for being trusted by an agent a_j that a_i distrusts. Obviously, this assumption is not sound and does not reflect expected real-world behavior. We assume that distrust does not allow for making direct inferences *of any kind*. This conservative assumption well complies with [Guha, 2003].

7.5.2 Incorporating Distrust into Appleseed

We compare our distrust model with Guha's approach, making similar assumptions. Guha computes trust by means of *one global* group trust metric, similar to PageRank [Page et al, 1998]. For distrust, he proposes two candidate approaches. The first one directly integrates distrust into the iterative eigenvector computation and comes up with one single measure combining both trust and distrust. However, in networks

---

[9] We oversimplify by using predicate calculus expressions, supposing that trust, and hence distrust, is fully transitive.

dominated by distrust, the iteration might not converge [Guha, 2003]. The second proposal first computes trust ranks by trying to find the dominant eigenvector, and then computes separate distrust ranks in one single step, based upon the iterative computation of trust ranks. Suppose that D_{a_i} is the set of agents who distrust a_i:

$$\text{DistrustRank}(a_i) = \frac{\sum_{a_j \in D_{a_i}} \text{TrustRank}(a_j)}{|D_{a_i}|} \tag{7.13}$$

The problem we perceive with this approach refers to *superimposing* the computation of distrust ranks *after* trust rank computation, which may yield some strange behavior: suppose an agent a_i who is highly controversial by engendering ambiguous sentiments, i.e., on the one hand, there are numerous agents that *trust* a_i, while on the other hand, there are numerous agents who *distrust* a_i. With the approach proposed by Guha, a_i's impact for distrusting other agents is huge, resulting from his immense positive trust rank. However, this should clearly not be the case, for a_i is subject to tremendous distrust himself, thus levelling out his high trust rank.

Hence, for our own approach, we intend to *directly* incorporate distrust into the iterative process of the Appleseed trust metric computation, and not superimpose distrust afterwards. Several pitfalls have to be avoided, such as the risk of non-convergence in case of networks dominated by distrust [Guha, 2003]. Furthermore, in absence of distrust statements, we want the distrust-enhanced Appleseed algorithm, which we denote by Trust_{α^-}, to yield results identical to those engendered by the original version Trust_{α}.

7.5.2.1 Normalization and Distrust

First, the trust normalization procedure has to be adapted. We suppose normalization of weights to the power of q, as has been discussed in Section 7.4.2.6. Let $\text{in}(x)$, the trust influx for agent x, be *positive*. As usual, we denote the global spreading factor by d, and quantified trust statements from x to y by $W(x,y)$. Function $\text{sign}(x)$ returns the sign of value x. Note that from now on, we assume $W : E \rightarrow [-1,+1]$, for degrees of *distrust* need to be expressible. Then the trust quantity $e_{x \rightarrow y}$ passed from x to successor y is computed as follows:

$$e_{x \rightarrow y} = d \cdot \text{in}(x) \cdot \text{sign}(W(x,y)) \cdot w, \tag{7.14}$$

where

$$w = \frac{|W(x,y)|^q}{\sum_{(x,s) \in E} |W(x,s)|^q}$$

The accorded quantity $e_{x \rightarrow y}$ becomes *negative* if $W(x,y)$ is negative, i.e., if x distrusts y. For the relative weighting, the *absolute* values $|W(x,s)|$ of all weights are considered. Otherwise, the denominator could become negative, or positive trust statements could become boosted unduly. The latter would be the case if the sum of positive trust ratings *only slightly* outweighed the sum of negative ones, making

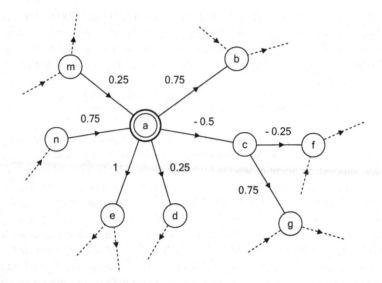

Fig. 7.9 Network augmented by distrust

the denominator converge towards zero. An example demonstrates the computation process:

Example 5 (Distribution of Trust and Distrust). We assume the trust network as depicted in Figure 7.9. Let the trust energy influx into node a be $in(a) = 2$, and global spreading factor $d = 0.85$. For simplicity reasons, backward propagation of trust to the source is *not* considered. Moreover, we suppose *linear* weight normalization, thus $q = 1$. Consequently, the denominator of the normalization equation is $|0.75| + |-0.5| + |0.25| + |1| = 2.5$. The trust energy that a distributes to b hence amounts to $e_{a \to b} = 0.51$, whereas the energy accorded to the distrusted node c is $e_{a \to c} = -0.34$. Furthermore, we have $e_{a \to d} = 0.17$ and $e_{a \to e} = 0.68$.

Observe that trust energy becomes *lost* during distribution, for the sum of energy accorded along outgoing edges of a amounts to 1.02, while 1.7 was provided for distribution. The effect results from the negative trust weight $W(a, c) = -0.5$.

7.5.2.2 Distrust Allocation and Propagation

We now analyze the case where the influx $in(x)$ for agent x is *negative*. In this case, the trust allocated for x will also be negative, i.e., $in(x) \cdot (1 - d) < 0$. Moreover, the energy $in(x) \cdot d$ that x may distribute among successor nodes will be negative as well. The implications are those which have been mentioned in Section 7.5.1, i.e., distrust as negation of trust and propagation of distrust. For the first case, refer to node f in Figure 7.9 and assume $in(c) = -0.34$, which is derived from Example 5. The trusted agent a distrusts c who distrusts f. Eventually, f would be accorded $d \cdot$

$(-0.34) \cdot (-0.25)$, which is *positive*. For the second case, node g would be assigned the *negative* trust quantity $d \cdot (-0.34) \cdot (0.75)$, simply for being trusted by f, who is distrusted. Both unwanted effects can be avoided by not allowing distrusted nodes to distribute *any energy at all*. Hence, more formally, we introduce a novel function $\text{out}(x)$:

$$\text{out}(x) = \begin{cases} d \cdot \text{in}(x), & \text{if } \text{in}(x) \geq 0 \\ 0, & \text{else} \end{cases} \qquad (7.15)$$

This function then has to replace $d \cdot \text{in}(x)$ when computing the energy distributed along edges from x to successor nodes y:

$$e_{x \to y} = \text{out}(x) \cdot \text{sign}(W(x,y)) \cdot w, \qquad (7.16)$$

where

$$w = \frac{|W(x,y)|^q}{\sum_{(x,s) \in E} |W(x,s)|^q}$$

This design decision perfectly aligns with assumptions made in Section 7.5.1 and prevents the inference of unwanted side-effects mentioned before. Furthermore, one can see easily that the modifications introduced *do not affect* the behavior of Algorithm 7.3 when not considering relationships of distrust.

7.5.2.3 Convergence

In networks largely or entirely dominated by distrust, the extended version of Appleseed is still guaranteed to converge. We therefore briefly outline an informal proof, based on Proof 7.4.2.7:

Proof 1 (Convergence in presence of distrust). Recall that only *positive* trust influx $\text{in}(x)$ becomes propagated, which has been indicated in Section 7.5.2.2. Hence, all we need to show is that the overall quantity of *positive* trust distributed in computation step i cannot be augmented through the presence of distrust statements. In other words, suppose that $G = (A, E, W)$ defines an arbitrary trust graph, containing quantified trust statements, but *no distrust*, i.e., $W : E \to [0,1]$. Now consider another trust graph $G' = (A, E \cup D, W')$, which contains additional edges D, and weight function $W' = W \cup (D \to [-1,0[)$. Hence, G' augments G by additional distrust edges between nodes taken from A. We now perform two parallel computations with the extended version of Appleseed, one operating on G and the other on G'. In every step, and for every trust edge $(x,y) \in E$ for G, the distributed energy $e_{x \to y}$ is greater or equal to the respective counterpart on G', because the denominator of the fraction given in Equation 7.16 can only become *greater* through additional distrust outedges. Second, for the computation performed on G', negative energy distributed along edge (x,y) can only *reduce* the trust influx for y and may hence even accelerate convergence.

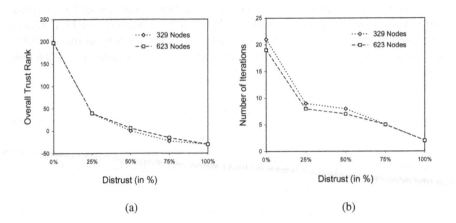

Fig. 7.10 Network impact of distrust

However, as can be observed from the proof, there exists one serious implication arising from having distrust statements in the network: the overall accorded trust quantity does *not* equal the initially injected energy anymore. Moreover, in networks dominated by distrust, the overall trust energy sum may even be *negative*.

Experiment 3 (Network impact of distrust). We observe the number of iterations until convergence is reached, and the overall accorded trust rank of 5 networks. The structures of all these graphs are identical, being composed of 623 nodes with an average indegree and outdegree of 9. The only difference applies to the assigned weights, where the first graph contains no distrust statements at all, while 25% of all weights are negative for the second, 50% for the third, and 75% for the fourth. The fifth graph contains nothing but distrust statements. The Appleseed parameters are identical for all 5 runs, having backward propagation enabled, an initial trust injection $in^0 = 200$, spreading factor $d = 0.85$, convergence threshold $T_c = 0.01$, *linear* weight normalization, and no upper bound on the number of nodes to unfold. Figure 7.10(a) clearly demonstrates that the number of iterations until convergence, given on the vertical axis, *decreases* with the proportion of distrust increasing, observable along the horizontal axis. Likewise, the overall accorded trust rank, indicated on the vertical axis of Figure 7.10(b), decreases rapidly with increasing distrust, eventually dropping below zero. The same experiment was repeated for another network with 329 nodes, an average indegree and outdegree of 6, yielding similar results.

The effects observable in Experiment 3 only marginally affect the ranking itself, for trust ranks are interpreted *relative* to each other. Moreover, compensation for lost trust energy may be achieved by boosting the initial trust injection in^0.

7.6 Discussion and Outlook

We provided a new classification scheme for trust metrics along three non-orthogonal feature axes. Moreover, we advocated the need for local group trust metrics, eventually presenting Appleseed, this chapter's main contribution. Appleseed's nature largely resembles Advogato, bearing similar complexity and attack-resistance properties, but offers one particular feature that makes Appleseed much more suitable for certain applications than Advogato: the ability to compute *rankings* of peers according to their trustworthiness rather than *binary* classifications into trusted and untrusted agents.

Originally designed as an approach to social filtering within our decentralized recommender framework, Appleseed suits other application scenarios as well, such as group trust computation in online communities, open rating systems, ad-hoc and peer-to-peer networks:

For instance, Appleseed could support peer-to-peer-based file-sharing systems in reducing the spread of self-replicating inauthentic files by virtue of trust propagation [Kamvar et al, 2003]. In that case, explicit trust statements, resulting from direct interaction, would reflect belief in someone's endeavor to provide authentic files.

We strongly believe that local group trust metrics, such as Advogato and Appleseed, will become subject to substantial research for diverse computing domains within the near future, owing to their favorable time complexity and their intuitive computation scheme, as opposed to other classes of trust metrics (see Section 7.3.2.2). However, one has to bear in mind that their applicability is confined to particular problem domains only, whereas scalar metrics are more versatile.

Chapter 8
Interpersonal Trust and Similarity

"Only your friends steal your books."

– Voltaire (1694 – 1778)

8.1 Introduction

Recently, the integration of computational trust models [Marsh, 1994b; Mui et al, 2002; McKnight and Chervany, 1996] into recommender systems has started gaining momentum [Montaner et al, 2002; Kinateder and Rothermel, 2003; Guha, 2003; Massa and Bhattacharjee, 2004], synthesizing recommendations based upon opinions from *most trusted* peers rather than *most similar*[1] ones. Likewise, for social filtering within a spread out and decentralized recommender framework, we cannot rely upon conventional collaborative filtering methods only, owing to the neighborhood computation scheme's poor scalability. Some more natural and, most important, *scalable* neighborhood selection process schemes become indispensable, e.g., based on trust networks.

However, in order to provide *meaningful* results, one should suppose trust to reflect user similarity to some extent. Clearly, recommendations only make sense when obtained from like-minded people having similar taste. For instance, Abdul-Rahman and Hailes [2000] claim that given some predefined domain and context, e.g., communities of people reading books, its members commence creating ties of friendship and trust primarily with persons resembling their *own* profile of interest. Jensen et al [2002] make likewise assumptions, supposing similarity as a strong predictor of friendship: "If I am a classic car enthusiast, for example, my friends will likely share my interests [...]. In other words, my circle of friends is likely to either share the same values as I do, or at least tolerate them."

Reasons for that phenomenon are manifold and mostly sociologically motivated, like people's striving for some sort of social affiliation [Given, 2002]. For instance,

---

[1] Requiring the explicit application of some similarity measure.

C.-N. Ziegler: *Social Web Artifacts for Boosting Recommenders*, SCI 487, pp. 133–151.
DOI: 10.1007/978-3-319-00527-0_8   © Springer International Publishing Switzerland 2013

Pescovitz [2003] describes endeavors to identify trust networks for crime preven-
tion and security. Hereby, its advocates operate "on the assumption that birds of a
feather tend to flock together [...]", an ancient and widely-known aphorism. However, though belief in the positive relation of trust and user similarity has been widely
adopted and presupposed, thus constituting the foundations for trust-based recommender and rating systems, to our best knowledge, no endeavors have been made
until now to provide "real-world" empirical evidence.

Hence, we want to investigate and analyze whether positive correlation actually holds, relying upon data mined from two online communities, one focusing
on books (which has been introduced before in Section 3.4.1), the other on movies.
Studies involve several hundreds of members telling which books or movies they
like and which other community members they trust. Our motivation derives from
incorporating trust models into recommender systems in order to reduce the complexity of neighborhood formation, which is the computational bottleneck of all
collaborative recommender systems. In this vein, trust shall not only be exploited
for selecting neighbors upon which to perform collaborative filtering, but also for
weeding out irrelevant peers up front.

However, before presenting our framework for conducting trust-similarity correlation experiments, we provide an outline of recommender systems employing trust
models, and an extensive survey giving results from socio-psychological research
that bear some significant relevance to our analysis.

8.2   Trust Models in Recommender Systems

Sinha and Swearingen [2001] have found that people prefer receiving recommendations from people they *know* and *trust*, i.e., friends and family-members, rather than
from online recommender systems. Some researchers have therefore commenced to
focus on computational trust models as appropriate means to supplement or replace
current collaborative filtering approaches.

Kautz et al [1997] mine social network structures in order to render expertise information exchange and collaboration feasible. Olsson [1998] proposes an architecture combining trust, collaborative filtering, and content-based filtering in one single
framework, giving only vague information and insight, though. Another agent-based
approach has been presented by Montaner et al [2002], who introduce the so-called
"opinion-based" filtering. Montaner claims that trust should be *derived* from user
similarity, implying that friends are exactly those people that resemble our very
nature. However, Montaner's model only extends to the agent world and does not
reflect evidence acquired from real-world social studies concerning the formation of
trust. Similar agent-based systems have been devised by Kinateder and Rothermel
[2003], Kinateder and Pearson [2003], and Chen and Singh [2001].

Apart from research in agent systems, online communities have also discovered
opportunities through trust network leverage. Epinions (*http://www.epinions.com*)
provides information filtering facilities based upon personalized webs of trust
[Guha, 2003]. Hereby, Guha states that the trust-based filtering approach has been

greatly approved and appreciated by Epinions' members. However, empirical and statistical justifications underpinning these findings, like indications of a correlation between trust and interest similarity, have not been part of Guha's work. Likewise, Massa and Avesani [2004] operate on Epinions and propose superseding CF-based neighborhoods by trust networks, making use of very basic propagation schemes. Initial empirical data has been provided in their work, indicating that precision does not decrease too much when using trust-based neighborhood formation schemes instead of common CF.

Besides the product rating portal Epinions, the All Consuming book readers' community (*http://www.allconsuming.net*, see Section 3.4.1), and the movie lovers platform FilmTrust (*trust.mindswap.org/FilmTrust/*, see Section 8.5) represent two further communities combining consumer ratings and trust networks.

8.3 Evidence from Social Psychology

Research in social psychology offers some important results for investigating interactions between trust and similarity. However, most relevant studies primarily focus on *interpersonal attraction* rather than trust, and its possible coupling with similarity. Interpersonal attraction constitutes a major field of interest of social psychology, and the positive impact of attitudinal similarity on liking has effectively become one of its most reliable findings [Berscheid, 1998]. Studies have given extensive attention to three different types of interpersonal relationships, namely same-sex friendships, primarily among college students, cross-sex romantic relationships, again primarily among college students, and marriage [Huston and Levinger, 1978]. Clearly, these three types of relationships also happen to be essential components of trust, though perfect equivalence does not hold. For instance, while friendship usually implies mutual trust, marriage does not. Moreover, the complex notion of interpersonal trust, already difficult to capture regarding the "lack of consensus" which has been pointed out by McKnight and Chervany [1996], interacts with other sociological concepts not reflected through interpersonal attraction. These elusive components comprise reputation, skill, situational and dispositional aspects of interpersonal trust [Marsh, 1994a,b], and familiarity [Einwiller, 2003].

However, since explicit *trust* relationships have remained outside the scope of empirical analysis on the correlation with attitudinal similarity, we are forced to stick to *interpersonal* attraction instead. Clearly, results obtained must be treated with care before attributing them to interpersonal trust as well. The following paragraphs hence intend to briefly summarize relevant evidence collected from research on interpersonal attraction.

8.3.1 On Interpersonal Attraction and Similarity

Early investigations date back until 1943, when Burgess and Wallin published their work about homogamy of social attributes with respect to engaged couples [Burgess and Wallin, 1943]. Similarity could be established for nearly every characteristic examined. However, according to Berscheid [1998], these findings do not justify conclusions about positive effects of similarity on interpersonal attraction by themselves, since "part of the association between similarity and social choice undoubtedly is due not to personal preference, but to the fact that people tend to be thrown together in time and space with others similar to themselves".

First large-scale experimental studies were conducted by Newcomb [1961] and Byrne [1961, 1971]. The former focused on friendships between American college students and nowadays counts among the seminal works on friendship formation. By means of his longitudinal study, Newcomb could reveal a positive association between attraction and attitudinal value similarity. Byrne, doing extensive research and experiments in the area of attraction, conducted similar experiments, applying the now-famous "bogus stranger technique" [Byrne, 1971]. The following section roughly outlines the original setup of this technique.

8.3.1.1   The Bogus Stranger Technique

First, all participating subjects had to indicate their preference on 26 diverse topics, ranging from more important ones, e.g., belief in super-natural beings, premarital sex, etc., to less important ones, like television programs, music taste, and so forth. Preference was expressed through 7-point likert scales. Two weeks later, the participants were falsely informed that they were now in a study on how well people can predict other people's behavior. In order to make these predictions, they were told that they would be given the attitude scales filled out by another participant. However, this was an outright lie. Actually, the scales were prepared by the experimenter, i.e., Byrnes and his assistants, in such way as to either reflect *high similarity* or *dissimilarity* with the subject's own profile. Participants were asked some questions thereafter about the respective "other participant", including personal sentiments toward the person and how much they would appreciate working with him. Moreover, participants were requested to evaluate the "bogus stranger" with respect to his intelligence, knowledge of current events, morality, and adjustment.

8.3.1.2   Analysis of Similarity-Attraction Associations

The result of Byrne's bogus stranger experiment well aligned with Newcomb's findings and confirmed that attitude similarity is a determinant of attraction. Rather than further document this fact, which counts among the most reliable findings in social psychology today [Berscheid, 1998], researchers have ever since attempted to identify the factors that mediate and define the *limitations* of positive association between similarity and attraction.

For instance, along with other theorists, e.g., Festinger and Newcomb, Byrne conjectured that one's mere discovery of some other person holding similar attitudes is reinforcing in itself, arguing that "the expression of similar attitudes by a stranger serves as a positive reinforcement, because consensual validation for an individual's attitudes and opinions and beliefs is a major source of reward for the drive to be logical, consistent, and accurate in interpreting the stimulus world" [Byrne, 1971].

We suppose likewise effects when forging bonds of trust. Hence, the sheer observation that some other peer holds interests similar to our own, e.g., reading the same kinds of books, intuitively renders the latter more trustworthy in our eyes and engenders sentiments of "familiarity". In fact, automated collaborative filtering systems exploit this conjecture in order to make reliable predictions of product preference [Sinha and Swearingen, 2001].

Social psychologists have identified some other likely factors accounting for the similarity-attraction association. For example, the information that another person possesses similar attitudes may suggest his sympathy towards the individual, and "it is known that the anticipation of being liked often generates attraction in return" [Berscheid, 1998]. Jones et al [1972] provided some large-scale empirical analysis for reciprocation of attraction from similar others.

8.3.1.3 Limitations

While positive association was attested for attitudinal similarity and interpersonal attraction, evidence could not be expanded to similarity in general. Berscheid [1998] therefore notes that despite "considerable effort to find a relationship between friendship choice and personality (as opposed to attitude) similarity, for example, the evidence for this hypothesis remains unconvincing [...]".

This inability to establish an association between personality similarity and attraction does not prove overly harmful to our hypothesis, since personal interests represent traits of *attitude* rather than *personality*. However, even attitude similarity fails to produce attraction under certain circumstances. Snyder and Fromkin [1980] reveal that perceiving very high similarity with another individual may even evoke *negative* sentiments towards that respective person. Moreover, according to Heider [1958], "similarity can evoke disliking when the similarity carries with it disagreeable implications", which common sense anticipates, likewise. Take narcissist persons as an example.

8.3.2 Conclusion

The preceding literature survey has shown that interactions between *similarity* traits and *interpersonal attraction* are difficult to capture. Even though the tight coupling between both concepts counts among social psychology's most reliable findings, there are numerous caveats to take into consideration, like subtle distinctions between various types of similarity, e.g., attitudinal similarity and personality similarity. Moreover, most studies argue that attitudinal similarity *implies* attraction, whereas the latter proposition's inversion, i.e., positing that similarity *follows* from attraction, has been subject to sparse research only. Common sense supports this hypothesis, though, since people tend to adopt attitudes of friends, spouses, etc.

8.4 Trust-Similarity Correlation Analysis

Even when taking reciprocal action between attitudinal similarity, and hence similarity of interests, and interpersonal attraction for granted, evidence from socio-psychological research does *not* provide sufficient support for positive interactions between *trust* and *interest similarity*. Mind that trust and interpersonal attraction, though subsuming several common aspects, e.g., friendship, familiarity, etc., are *not* fully compliant notions.

We hence intend to establish a formal framework for investigating interactions between trust and similarity, believing that given an application domain, such as, for instance, the book-reading domain, people's trusted peers are on average considerably more similar to their sources of trust than arbitrary peers. More formally, let A denote the set of all community members, $\text{trust}(a_i)$ the set of all users trusted by a_i, and $\text{sim} : A \times A \rightarrow [-1, +1]$ some similarity function:

$$\sum_{a_i \in A} \frac{\sum_{a_j \in \text{trust}(a_i)} \text{sim}(a_i, a_j)}{|\text{trust}(a_i)|} \gg \sum_{a_i \in A} \frac{\sum_{a_j \in A \setminus \{a_i\}} \text{sim}(a_i, a_j)}{|A| - 1} \qquad (8.1)$$

For instance, given that agent a_i is interested in Science-Fiction and Artificial Intelligence, chances that a_j, trusted by a_i, also likes these two topics are much higher than for peer a_e not explicitly trusted by a_i. Various social processes are involved, such as participation in those social groups that best reflect our own interests and desires.

8.4.1 Model and Data Acquisition

In order to verify or refute our hypothesis for some specific domain, we need to define an information model, determine metrics and methods for evaluation, and apply our framework to real-world data.

8.4.1.1 Information Model

We assume the same model as the one presented in Section 3.3.1, but provide an extension for trust networks. Function $\text{trust} : A \rightarrow 2^A$ gives the set of all users that agent a_i trusts. Hence, for the scenario at hand, we assume trust as a relationship of *binary* preference, dividing the set of agents A into trusted and non-trusted ones for every $a_i \in A$. For user-user similarity computations $c(a_i, a_j) : A \times A \rightarrow [-1, +1]$, we employ our taxonomy-driven metric, presented in Section 3.3. Its huge advantage over pure CF similarity measures (see Section 2.3.2) lies in its ability to also work for *sparse* domains: when two users have no overlap in their purchased or rated products, pure CF measures become unable to make any reasonable inferences with respect to interest similarity, which is *not* the case for the taxonomy-driven method. Since All Consuming, the dataset we conduct all experiments upon, offers comparatively few ratings taken from a large product set, the ability to handle sparsity becomes an indispensable feature for eligible similarity metrics.

The following section now relates our supposed information model to an actual scenario, making use of variable and function bindings introduced above.

8.4.1.2 Data Acquisition

All Consuming is one of the few communities that allow members to express which other users they trust, as well as which items, in our case books, they appreciate. Hereby, users may *import* their list of trusted persons from other applications like FOAF [Dumbill, 2002]. All Consuming also offers to automatically compile information about books its members have read from their personal weblog. In contrast to the All Consuming dataset described in Chapter 3, the one used in this chapter has been crawled earlier, from October 13 to October 17, 2003.

Our tools mined data from about 2,074 weblogs contributing to the All Consuming information base, and 527 users issuing 4.93 trust statements on average. These users have mentioned 6,592 different books altogether. In order to obtain category descriptors $f(b_k)$ for all discovered books b_k, classification information from the Amazon.com online shop (*http://www.amazon.com*) was garnered. For each book, we collected an average of about 4.1 classification topics, relating books to Amazon.com's book taxonomy.

8.4.2 Experiment Setup and Analysis

This section describes the two experiments we performed in order to analyze possible positive correlations between interest similarity and interpersonal trust. In both cases, experiments were run on data obtained from All Consuming (see Section 8.4.1.2). Considering the slightly different information makeup the two experiments were based upon, we expected the first to define an *upper bound* for the analysis, and the second one a *lower bound*. Results obtained confirmed our assumption.

8.4.2.1 Upper Bound Analysis

Before conducting the two experiments, we applied extensive data cleansing and duplicate removal to the All Consuming *active user* base of 527 members[2]. First, we pruned all users a_i having fewer than 3 books mentioned, removing them from user base A and from all sets trust(a_j) where $a_i \in$ trust(a_j). Next, we discarded all users a_i who did not issue any trust assertions at all. Interestingly, some users had created *several* accounts. We discovered these "duplicates" by virtue of scanning through account names for similarity patterns and by tracking identical or highly similar profiles in terms of book mentions. Moreover, we stripped self-references, i.e., statements about users trusting themselves. Through application of data cleansing, 266 users were discarded from the initial test set, leaving 261 users for the

---

[2] All Consuming's crawled weblogs were *not* considered for the experiments, owing to their lack of trust web information.

upper bound experiment to run upon. We denote the reduced set of users by A' and corresponding trust functions by $\text{trust}'(a_i)$.

For every single user $a_i \in A'$, we generated his profile vector and computed the similarity score $c(a_i, a_j)$ for each *trusted* peer $a_j \in \text{trust}'(a_i)$. Next, we averaged these proximity measures, obtaining value z'_i:

$$z'_i := \frac{\sum_{a_j \in \text{trust}'(a_i)} c(a_i, a_j)}{|\text{trust}'(a_i)|} \tag{8.2}$$

Moreover, we computed a_i's similarity with any other user from dataset A', except a_i himself. Again, we took the average of these proximity measures and recorded the result s'_i:

$$s'_i := \frac{\sum_{a_j \in A' \setminus \{a_i\}} c(a_i, a_j)}{|A'| - 1} \tag{8.3}$$

A comparison of pairs (z'_i, s'_i) revealed that in 173 cases, users were more similar to their trusted peers than arbitrary ones. The opposite held for only 88 agents. Users had an average similarity score of 0.247 with respect to their trusted peers, while only exhibiting 0.163 with complete A'. In other words, users were *more than 50%* more similar to their trusted agents than arbitrary peers.

Distributions of z' and s'

Figure 8.1 gives histogram representations for z' and s', respectively. No agents have higher average similarity than 0.4, i.e., $s'_i \leq 0.4$ holds for all $a_i \in A'$. This is not the case for z', as there remains a considerable amount of users a_i exhibiting an average trusted-peer similarity z'_i larger than 0.4. About 20 agents have $z'_i > 0.6$. Interestingly, while the overall peer similarity s' shows an almost perfect Gaussian distribution curve, its counterpart z' does not feature the typical bell shape. This observation raises some serious concerns when conducting analysis of statistical significance in Section 8.4.3.

Scatter Plot

In order to directly match every user's overall similarity s'_i against his average trusted-peer similarity z'_i, Figure 8.2 provides a scatter plot for the experiment at hand. The dashed line, dividing the scheme into an upper and lower region, models an agent a_i having *identical* similarity values, i.e., $s'_i = z'_i$. Clearly, the plot exhibits a strong bias towards the upper region, which becomes particularly pronounced for agents a_i with $s'_i > 0.15$.

8.4.2.2 Lower Bound Analysis

The first experiment proposed that users tend to trust people that are significantly more similar to themselves than average users. However, we have to consider that

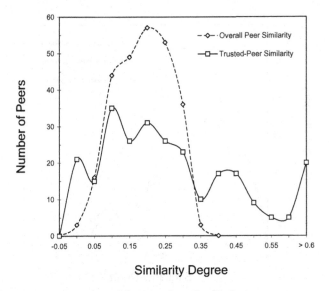

Fig. 8.1 Histogram representation of the upper bound analysis

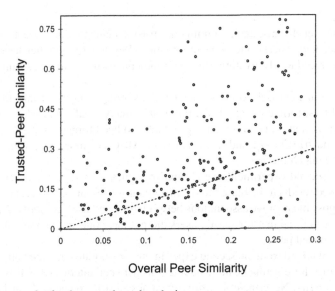

Fig. 8.2 Scatter plot for the upper bound analysis

All Consuming offers a feature that *suggests friends* to newbie users a_i. Hereby, All Consuming chooses users who have *at least one book in common* with a_i. Hence, we have reason to suspect that our first experiment was biased and too optimistic with respect to positive interactions between trust and similarity. Consequently, we

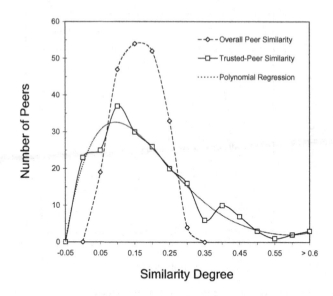

Fig. 8.3 Histogram representation of the lower bound analysis

pruned user set A' once again, eliminating trust statements whenever trusting and trusted user had at least one book in common. We denote the latter user base by A'', now reduced to 210 trusting users, and indicate its respective trust functions by $trust''(a_i)$.

Clearly, our approach to eliminate All Consuming's intrusion into the natural process of trust formation entails the removal of many "real" trust relationships between users a_i and a_j, i.e., relationships which had been forged owing to a_i actually *knowing* and *trusting* a_j, and not because of All Consuming proposing a_j as an appropriate match for a_i.

For the second experiment, we computed values s_i'' and z_i'' for every $a_i \in A''$. We supposed results to be biased to the *disadvantage* of our conjecture, i.e., unduly lowering possible positive associations between trust and user similarity. Again, one should bear in mind that for set A'', users did not have one single book in common with their trusted peers.

Results obtained from the second experiment corroborate our expectations, being less indicative for existing positive interactions between interpersonal trust and attitudinal similarity. Nevertheless, similarity values z_i'' still exceeded s_i'': in 112 cases, people were more similar to their trusted fellows than arbitrary peers. The opposite held for 98 users. Mean values of z'' and s'' amounted to 0.164 and 0.134, respectively. Hence, even for the lower bound experiment, users were still approximately 23% more similar to their trusted fellows than arbitrary agents.

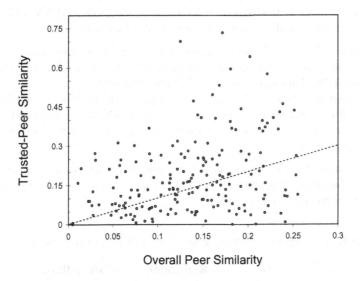

Fig. 8.4 Scatter plot for the lower bound analysis

Histogram Curves

The bell-shaped distribution of s'', depicted in Figure 8.3, looks more condensed with respect to s' and has its peak slightly below the latter plot's curve. The differences between z'' and z' are even more pronounced, though, e.g., the shape of z'''s histogram looks more "regular" than z''s pendant. Hence, the approximation of z'''s distribution, applying polynomial regression of degree 5, strongly resembles the Erlang-k distribution, supposing $k = 2$. For similarity degrees above 0.35, peaks of z'''s histogram are considerably less explicit than for z'' or have effectively disappeared, as is the case for degrees above 0.6.

Matching z_i'' Against s_i''

Figure 8.4 gives the scatter plot of our lower bound analysis. The strong bias towards the upper region has become less articulate, though still clearly visible. Interestingly, the increase of ratio $z_i'' : s_i''$ for $s_i'' > 0.15$ still persists.

8.4.3 Statistical Significance

We conclude our experimental analysis noticing that without exact knowledge of how much noise All Consuming's "friend recommender" adds to our obtained results, we expect the *true* correlation intensity between trust and interest similarity to reside somewhere within our computed upper and lower bound.

Moreover, we investigated whether the increase of mean values of z' with respect to s', and z'' with respect to s'', bears statistical significance or not.

For the analyses at hand, common parametrical one-factor ANOVA could *not* be applied to z' and s', and z'' and s'', likewise, for diverse reasons:

Gaussian distribution. The distributions of both samples have to be *normal*, even though small departures may be accommodated. While s' and s'' exhibit the latter Gaussian distribution property, z' and z'' obviously do not.

Equal variances. Data transformation, e.g., logarithmic, probits, etc., might be an option for z'', bearing traits of Erlang-2. However, ANOVA also demands largely identical *variances* σ^2. Since z'''s variance is 5.33 times the variance of s'', this criterion cannot be satisfied.

Hence, owing to these two limitations, we opted for Kruskal-Wallis non-parametric ANOVA [Siegel and Castellan, 1988], which does not make any assumptions with respect to distribution and variance.

Table 8.1 Kruskal-Wallis ANOVA test results for the upper bound experiment

	n	**Rank Sum**	**Mean Rank**
z'	261	73702.0	284.56
s'	261	60719.0	234.44
	Kruskal-Wallis Statistic		14.52
	p		**0.0001**

Table 8.1 shows result parameters obtained from analyzing the upper bound experiment. Since value p is much smaller than 0.05, very high statistical significance holds, thus refuting the hypothesis that fluctuations between medians of s' and z' were caused by mere random.

For the lower bound experiment, on the other hand, no statistical significance was detected, indicated by large p and a low Kruskal-Wallis statistic being much smaller than 1 (see Table 8.2).

8.4.4 Conclusion

Both experiments suggest that the mean similarity of trusting and trusted peers exceeds the arbitrary user similarity. For the upper bound analysis, strong statistical significance was discovered, which was not the case for its lower bound pendant. However, assuming the true distribution curves to reside somewhere in between these bounds, and taking into account that both z' and z'' exhibit larger mean values than s' and s'', respectively, the results we obtained bear strong indications towards positive interactions between interpersonal trust and interest similarity.

Table 8.2 Kruskal-Wallis ANOVA test results for the lower bound experiment

	n	**Rank Sum**	**Mean Rank**
z''	210	43685.0	210.02
s''	210	43051.0	206.98

Kruskal-Wallis Statistic	0.07
p	**0.796**

8.5 Trust and Similarity in FilmTrust

FilmTrust (*http://trust.mindswap.org/FilmTrust/*) is a Web site that integrates social networks with movie ratings and reviews [Golbeck, 2005]. Users make connections to friends, and, more importantly for this work, rate how much they trust each friend's opinion about movies. Users can also rate films on a scale from one half star (very bad) to four stars (excellent). Within the site, the social network is used as a recommender system to generate *predictive* ratings for movies based on the user's trust values for others in the network. The data from this Web site also provides an opportunity to study the relationship of user similarity and trust in a context where both trust values and opinions about movies have been *explicitly* rated on a scale. The section at hand will introduce the FilmTrust network and use the data to illustrate a strong and significant correlation between trust and user similarity.

Note that by virtue of the different, prediction-oriented information model supposed for FilmTrust, the design of the study substantially differs from the evaluation framework presented in previous sections, thus complementing our latter approach.

8.5.1 FilmTrust Introduction

The social networking component of the FilmTrust Web site allows users to make connections to friends. There are currently[3] over 400 members of the FilmTrust community. Figure 8.5 shows the structure of the social network. When adding a friend, users are required to provide a trust rating for that person. Because the context of the Web site is movie-specific, users are asked to rate how much they trust their friends' opinions of movies. Ratings are made on a 1 through 10 scale, where 1 represents very low trust and 10 very high trust. There is no rating to reflect distrust in the system because the meaning of distrust is far less clear, both socially and computationally.

---

[3] User numbers as of November 2006.

Fig. 8.5 Visualization of the FilmTrust network's largest connected component

Part of the user's profile is a "Friends" page. In the FilmTrust network, relationships can be one-way, so the page displays a list of people the user has named as friends, and a second list of people who have named the user as a friend. An icon indicates reciprocal relationships and the trust ratings that the user assigned are shown next to each friend.

If trust ratings are visible to everyone, users can be discouraged from giving accurate ratings for fear of offending or upsetting people by giving them low ratings. Because honest trust ratings are important to the function of the system, these values are kept private and shown only to the user who assigned them. The ratings that people assigned to the user are not shown.

The other relevant feature of the Web site is a movie rating and review system. Users can choose any film and rate it on a scale of one half star to four stars. They can also write free-text reviews about movies. The user's "Movies" page displays data for every movie that he has rated or reviewed.

These explicit ratings of trust values and movies facilitate an analysis of the correlation between trust and user similarity.

8.5.2 Profile Similarity Computation

When users join the FilmTrust network, they are presented with a list of the top 50 films on the American Film Institute's top 100 movies list[4] and asked to assign a star rating to any movies they have seen. Thus, there is a core set of films with ratings from many users that permit a better and more facile analysis.

Let $t_i(a_j)$ represent the trust rating that user $a_i \in A$ has assigned to user a_j and $r_i(b_k)$ represent the rating user a_i has assigned to movie b_k. Each neighbor of user a_i is contained in an adjacency list given by $\text{adj}(a_i)$. Each movie rated by user a_i is contained in the set $R_i = \{b \in B \,|\, r_i(b) \neq \perp\}$.

Similarity between users a_i and a_j on a given movie b_k is computed as the absolute difference $\delta = |r_i(b_k) - r_j(b_k)|$. To compute the correlation between trust and similarity, neighbors are grouped according to trust value. Since trust is allocated on an integer scale from 1 to 10, there are ten groups of neighbors. Let τ represent the trust value for which the similarity is being computed. Each movie that has been rated by user a_i and user a_j where $t_i(a_j) = \tau$ is compared using the absolute difference δ. The average δ, $\bar{\delta}$, is used as the measure of similarity for trust value τ. That is, we want the average $|r_i(b_k) - r_j(b_k)|$, $\forall a_i, a_j \in A : t_i(a_j) = \tau, \forall b_k \in R_i \cap R_j$.

For example, when $\tau = 9$, we select all user pairs where user a_i has rated user a_j with a trust value of 9. Then, we select all movies b_k rated by both user a_i and user a_j, and take the absolute difference of their ratings for each b_k. The average of these differences over all movies for all $(a_i, a_j) \in A \times A$ where $t_i(a_j) = 9$ is used as a measure of similarity for $\tau = 9$.

As shown in Figure 8.6, as the trust rating increases, $\bar{\delta}$ decreases. Recall that the difference in movie ratings is a measure of *error*, so a lower difference means higher similarity. Thus, we see that user similarity increases as trust increases.

8.5.3 Statistical Significance

There are two questions as to the significance of these results. First, is there a statistical correlation between trust and similarity, and second, is the difference statistically significant.

Figure 8.6 appears to show a *linear* relationship between $\bar{\delta}$ and trust. We compute this with the Pearson correlation [Shardanand and Maes, 1995]. It is important to point out that both trust rating and $\bar{\delta}$ values are on a scale of discrete values. Although the data are not continuous, they are not strictly categorical either. We choose to use the Pearson correlation, but because of the scalar nature of the values, the coefficient serves more as a general indicator than a precise measurement of the correlation.[5] Nevertheless, the correlation between trust and $\bar{\delta}$ is strong and significant. For these data, the correlation coefficient $r = -0.987$ indicates an almost perfect negative linear relationship: as trust increases, the average difference

---

[4] To obtain the complete list, refer to *http://www.afi.com/tvevents/100years/movies.aspx*.

[5] The analysis methodology presented herein resembles the approach adopted by Newman [2003] to investigate the strength of *assortative mixing* in social networks.

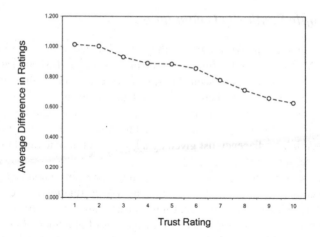

Fig. 8.6 Mean difference $\bar{\delta}$ grouped by trust rating

in ratings decreases (ratings become more similar). For these sample sizes, this correlation is significant at $p < 0.01$.

To test the significance of the change in $\bar{\delta}$ as trust values change, we used the same Kruskal-Wallis ANOVA as was used before in the All Consuming study. While the sample distributions are normal in this case, the variance is different in each group, making the standard ANOVA inappropriate. Table 8.3 shows the result parameters of the Kruskal-Wallis ANOVA. The p value is quite small, indicating that $\bar{\delta}$ decreases significantly as trust increases.

8.5.4 Conclusion

The data from the FilmTrust Web site give insight into how user similarity changes when there is a range of trust values. In the analysis presented here, we see that as the trust between users increases, the difference in the ratings they assign to movies decreases. This is a significant change, and the correlation between trust and similarity is strong. Our finding reinforces the results from Section 8.4 by showing that changes in similarity are correlated to changes in trust between users.

8.6 Exploiting Correlations between Trust and Similarity

Knowledge about positive correlation between trust and interest similarity may be exploited for diverse applications. In particular, we envision trust to play an important role for *decentralized* recommender systems. These filtering systems suppose distributed data and control and currently face various problems inherent to their very nature:

Table 8.3 Kruskal-Wallis ANOVA test results for FilmTrust data

Trust Value	n	Rank Sum	Mean Rank
1	158	636734.5	4029.97
2	203	820833.5	4043.51
3	195	752781.0	3860.42
4	296	1129921.0	3817.30
5	1235	4681499.5	3790.69
6	1043	3874654.5	3714.91
7	1217	4233673.5	3478.78
8	1142	3771554.0	3302.59
9	636	2049333.5	3222.22
10	898	2713791.0	3022.04

Kruskal-Wallis Statistic	17.96
p	**0.0001**

Credibility and attack-resistance. The Semantic Web and other open systems lack dedicated mechanisms and facilities to verify user identity. Hence, these systems tend to encourage insincerity and fraudulent behavior. Moreover, penalization and banishment are hard to accomplish and facile to short-circuit. Collaborative filtering becomes particularly susceptive to attack, for malicious users simply have to create profiles replicating the victim's in order to obtain high similarity. Then they can lure the victim into buying items the purchase of which may provide some utility for the attacker.

Product-user matrix sparseness. Communities often limit the number of ratable products, therefore avoiding product-user matrices from becoming overly sparse. Besides, Ringo [Shardanand and Maes, 1995] and other systems require users to rate items from *small product subsets* to generate user profiles with sufficient overlap. However, decentralized recommender system cannot suppose reduced item sets. Bear in mind that controlling product set contents and having users rate certain goods presupposes some central authority.

Computational complexity and scalability. Centralized systems are able to control and limit the number of members. Depending on the community's size, large-scale server clusters ensure proper operativeness and scalability. In general, recommender systems imply heavy computations. For instance, collaborative filtering systems compute Pearson correlation for users a_i offline rather than on-the-fly. Recall that coefficients $c(a_i, a_j)$ have to be computed for every other agent $a_j \in A$. Clearly, this approach does not work for large decentralized systems. Sensible prefiltering mechanisms which still ensure reasonable recall are needed.

Clearly, trust succeeds to address the credibility problem. Every agent builds his own neighborhood of trusted peers, relying upon direct trust statements and those from trusted peers, likewise. For deriving trust, numerous metrics have been proposed during the mid-nineties, among those [Maurer, 1996], [Abdul-Rahman and Hailes, 1997], [Beth et al, 1994], [Levien and Aiken, 1998], and more recently [Golbeck and Hendler, 2004; Golbeck et al, 2003]. However, we believe that local group trust metrics like Levien's ADVOGATO [Levien, 2004] and APPLESEED [Ziegler and Lausen, 2004c] best fit neighborhood formation in decentralized systems [Ziegler, 2004b]. Unfortunately, trust cannot handle product-user matrix sparseness, nor substantially reduce dimensionality. Supplementary approaches are needed, e.g., taxonomy-based filtering techniques [Ziegler et al, 2004a] similar to the one proposed.

Increased computational complexity and loss of scalability are mitigated and may even be eliminated when supposing positive correlation between trust and user similarity. Note that the complexity issue *per se* does not require the latter correlation to hold: limiting collaborative filtering to selected peers part of agent a_i's trust neighborhood entails complexity reduction, too. However, when supposing that trust does *not* reflect similarity, serious tradeoffs are implied, for scalability comes at the expense of neighborhood quality. Mind that trust neighborhood A_i of agent a_i only represents one tiny fraction of the overall system A. Moreover, this fraction does not necessarily contain similar peers. Instead, trusted agents are on average no more similar than arbitrary ones. Hence, the number of agents $a_j \in A_i$ with $c(a_i, a_j)$ above some threshold t, found by the filtering process, degrades *proportionally* with the neighborhood's size. On the other hand, when assuming that trust *does* correlate with similarity, the degradation does not take place as fast, thus ensuring reasonable neighborhood quality.

The approach pursued by Epinions [Guha, 2003; Guha et al, 2004] relies upon trust networks as only filtering mechanism, clearly exploiting the latter correlation. Positive user feedback backs the design decision. Nevertheless, we believe that trust should rather *supplement* than replace existing filtering techniques. For instance, ex-post application of collaborative filtering to computed trust neighborhoods A_i might boost precision significantly.

8.7 Discussion and Outlook

In this chapter, we articulated our hypothesis that positive mutual interactions between interpersonal trust and user similarity exist when the community's trust network is tightly bound to some particular application, e.g., reading books or watching movies.

Before presenting an evaluation framework to conduct empirical analyses, we provided an extensive literature survey on relevant socio-psychological research.

We then applied our evaluation method, using the taxonomy-driven similarity computation technique presented in Chapter 3, to the All Consuming community. In order to make our results even more robust, we conducted a second study, now on FilmTrust, a community catering to film lovers. The evaluation method was adapted to fit the scenario at hand. From this study, empirical evidence on positive correlation was backed. To our best knowledge, suchlike experiments have not been performed before, since communities incorporating explicit trust models are still very sparse.

We believe that our positive results will have substantial impact for ongoing research in recommender systems, where discovering similar users is of paramount importance. Decentralized approaches will especially benefit from trust network leverage. Hereby, the outstanding feature of trust networks lies in their ability to allow for sensible *pre*-filtering of like-minded peers and to increase the *credibility* of recommendations. Arbitrary social networks, on the other hand, only allow for reducing the computational complexity when composing neighborhoods, trading recommendation quality for scalability.

Though backing our experiments with information involving several hundreds of people, we believe that additional efforts studying trust-similarity interactions in domains other than books or movies are required in order to further corroborate our hypothesis. Unfortunately, at the time of this writing, the large-scale penetration of trust networks into communities, particularly those where users are given the opportunity to rate products, still has to take place.

Part IV
Amalgamating Taxonomies and Trust

Chapter 9
Decentralized Recommender Systems

"You must trust and believe in people or life becomes impossible."

– Anton Checkhov (1860 – 1904)

9.1 Introduction

Preceding chapters, particularly Chapter 3 and Chapter 7, have presented methods and techniques based on Web 2.0 information structures that are, among other things, able to address specific issues of *decentralized* recommender systems. Moreover, Chapter 8 has shown that, to a certain extent, trust *implies* similarity and thus becomes eligible as a tool for CF neighborhood formation, which is generally performed by applying some rating-based or attribute-based similarity measure (see Section 2.3.2).

In this chapter, we integrate the preceding contributions into one coherent, decentralized recommender framework. The presented architecture illustrates one sample approach, others are likewise conceivable and may represent equally appropriate solutions. Outstanding features of our method are the strong focus on *trust* and the usage of *taxonomy*-driven similarity measures.

Before delving into the details of our approach's mechanics, we will briefly introduce the research issues we see in particular with decentralized recommenders:

9.2 On Decentralized Recommender Systems

With few exceptions (see [Foner, 1999; Olsson, 2003; Miller, 2003; Sarwar, 2001]), recommender systems have been crafted with *centralized* scenarios in mind, i.e., central computational control and central data storage. On the other hand, *decentralized* infrastructures are becoming increasingly popular on the Internet and the Web (2.0), among those the Semantic Web, the Grid, peer-to-peer networks for file-sharing and collaborative tasks, and ubiquitous computing. All these

C.-N. Ziegler: *Social Web Artifacts for Boosting Recommenders*, SCI 487, pp. 155–172.
DOI: 10.1007/978-3-319-00527-0_9   © Springer International Publishing Switzerland 2013

scenarios comprise an abundant wealth of metadata information that could be exploited for personalized recommendation making.

For instance, think of the Semantic Web as an enormous network of inter-linked personal homepages published in a machine-readable fashion. All these homepages contain certain preference data, such as the respective user's friends and trusted peers, and appreciated products, e.g., CDs, DVDs, and so forth. Through weblogs, best described as online diaries, parts of this vision have already become true and gained wide-spread acceptance.

9.2.1   Decentralized Recommender Setup

We could exploit this existing information infrastructure, the user's personal preferences with respect to peers and products, in order to provide personalized recommendations of products he might have an interest in. This personal recommender would thus be an application running on one local node, namely the respective user's personal computer, and collect data from throughout the Semantic Web. Our devised recommender would thus exhibit the following characteristics:

Centralized computation. All computations are performed on one single node. Since we assume these nodes to be people's PCs, limitations are set concerning the computational power, i.e., we cannot suppose large clusters of high-speed servers as is the case for e-commerce stores and larger online communities.

Decentralized data storage. Though computations are localized, data and preference information are not. They are distributed throughout the network, e.g., the Semantic Web, and peers generally maintain partial views of their environment only.

Though having referred to the Semantic Web in the above example, the devised example also translates to other decentralized infrastructures, such as the beforementioned peer-to-peer file-sharing systems.

9.2.2   Research Issues

Now, when thinking of personal recommender systems exhibiting features as those depicted above, several research issues spring to mind that are either non-existent or less severe when dealing with conventional, centralized approaches:

- **Ontological commitment.** The Semantic Web and other decentralized infrastructures feature machine-readable content distributed all over the Web. In order to ensure that agents can understand and reason about the respective information, semantic interoperability via ontologies or common content models must be established. For instance, FOAF [Golbeck et al, 2003], an acronym for "Friend of a Friend", defines an ontology for establishing simple social networks and represents an open standard systems can rely upon.
- **Interaction facilities.** Decentralized recommender systems have primarily been subject to multi-agent research projects [Foner, 1997; Olsson, 1998; Chen and

Singh, 2001]. In these settings, environment models are *agent*-centric, enabling agents to directly communicate with their peers and thus making synchronous message exchange feasible. The Semantic Web, being an aggregation of distributed metadata, opts for an inherently *data*-centric environment model. Messages are exchanged by publishing or updating documents encoded in RDF, OWL, or similar formats. Hence, communication becomes restricted to asynchronous message exchange only.

- **Security and credibility.** Closed communities generally possess efficient means to control the users' identity and penalize malevolent behavior. Decentralized systems devoid of central authorities, e.g., peer-to-peer networks, open marketplaces and the Semantic Web, likewise, cannot prevent deception and insincerity. Spoofing and identity forging thus become facile to achieve [Lam and Riedl, 2004; O'Mahony et al, 2004]. Hence, some subjective means enabling each individual to decide which peers and content to rely upon are needed.

- **Computational complexity and scalability.** Centralized systems allow for estimating and limiting the community size and may thus tailor their filtering systems to ensure scalability. Note that user similarity assessment, which is an integral part of collaborative filtering [Goldberg et al, 1992], implies some computation-intensive processes. The Semantic Web will once contain millions of machine-readable homepages. Computing similarity measures for all these "individuals" thus becomes infeasible. Consequently, scalability can only be ensured when restricting these computations to sufficiently narrow neighborhoods. Intelligent filtering mechanisms are needed, still ensuring reasonable recall, i.e., not sacrificing too many relevant, like-minded agents.

- **Sparsity and low profile overlap.** As indicated in Section 1.1.1, interest profiles are generally represented by vectors showing the user's opinion for every product. In order to reduce dimensionality and ensure profile overlap, hence combatting the so-called sparsity issue [Sarwar, 2001], some centralized systems like MovieLens (*http://www.movielens.org*) and Ringo [Shardanand and Maes, 1995] prompt people to rate small subsets of the overall product space. These mandatory assessments, provisional tools for creating overlap-ensuring profiles, imply additional efforts for prospective users. Other recommenders, such as FilmTrust (see Section 8.5) and Jester [Goldberg et al, 2001], operate in domains where product sets are comparatively small. On the Semantic Web, virtually no restrictions can be imposed on agents regarding which items to rate. Instead, "anyone can say anything about anything", as stated by Tim Berners-Lee. Hence, new approaches to ensure profile overlap are needed in order to make profile similarity measures meaningful.

9.3 Principal Design Considerations

In order to address the research issues stated above, we propose a decentralized recommender architecture that revolves around trust networks and product classification taxonomies.

Hereby, trust networks allow for circumventing the complexity issue that those recommender systems operating in large decentralized settings are facing. Mind that similarity-based neighborhood computation impersonates a severe bottleneck, owing to the $O(|A|^2)$ time complexity when composing neighborhoods for all $|A|$ members part of the system. Hence, these approaches do not scale. On the other hand, network-based propagation approaches, e.g., Appleseed or Advogato (see Chapter 7), necessitate partial graph exploration only and scale to arbitrary network sizes.

Second, the low rating profile overlap issue [Sarwar et al, 2000b] that large communities with effectively unconstrained product sets are confronted with, investigated in detail by Massa and Bhattacharjee [2004] for the well-known Epinions community (*http://www.epinions.com*), is addressed through taxonomy-driven similarity computations (see Chapter 3).

Besides describing the make-up of our decentralized recommender framework, we conduct empirical analyses and compare results with benchmarks from two centralized architectures, namely one purely content-based system, and the taxonomy-driven filtering system proposed in Chapter 3.

Advocacy for Trust-based Neighborhood Formation

As stated above, we investigate social network structures in order to easily assemble personalized neighborhoods for active users a_i. To give an example of network-based neighborhood formation, a_i's neighborhood may comprise exactly those peers being closest in terms of *link distance*, necessitating simple breadth-first search instead of $O(|A|)$ complexity, which is required for computing similarity measures between one single a_i and all other individuals in the system. More specifically, we exclusively focus on *trust* relationships, motivated by reasons given below:

- **Security and attack-resistance.** Closed communities generally possess efficient means to control the user's identity and penalize malevolent behavior. However, decentralized systems cannot prevent deception and insincerity. Spoofing and identity forging thus become facile to achieve and allow for luring people into purchasing products which may provide some benefit for attackers a_o [Lam and Riedl, 2004; O'Mahony et al, 2004]. For instance, to accomplish such attacks, agents a_o simply have to copy victim a_v's rating profile and add excellent ratings for products b_k they want to trick a_v into buying. Owing to high similarities between rating profiles of a_o and a_v, b_k's probability of being proposed to a_v quickly soars beyond competing products' recommendation likelihood. On the other hand, only proposing products from people the active user deems most trustworthy inherently solves this issue, hence excluding perturbations from unknown and malevolent agents from the outset.
- **Recommendation transparency.** One of the major disadvantages of recommender systems relates to their lacking transparency, i.e., users would like to understand *why* they were recommended particular goods [Herlocker et al, 2000]. However, algorithmic clockworks of recommenders effectively resemble black boxes. Hence, when proposing products from users based upon complex similarity measures, most of these "neighbors" probably being unknown to the active

user, recommendations become difficult to follow. On the other hand, recommendations from trustworthy people clearly exhibit higher acceptance probability. Recall that trust metrics operate on naturally grown social network structures while neighborhoods based upon interest similarity represent pure artifacts, computed according to some obscure scheme.

- **Correlation of trust and similarity.** Sinha and Swearingen [2001] have found that people tend to prefer receiving recommendations from people they *know* and *trust*, i.e., friends and family-members, rather than from online recommender systems. Moreover, positive mutual impact of attitudinal similarity on interpersonal attraction counts among one of the most reliable findings of modern social psychology [Berscheid, 1998], backing the proverbial saying that "birds of a feather flock together". In Chapter 8, we have provided first empirical evidence confirming the positive correlation between trust and interest similarity.
- **Mitigating the new-user cold-start problem.** One major weakness that CF systems are faced with is the so-called new-user cold-start problem [Middleton et al, 2002]: newbie members generally have issued few product ratings only. Consequently, owing to common product-user matrix sparseness and low profile overlap, appropriate similarity-based neighbors are difficult to find, entailing poor recommendations. The whole process is self-destructive, for users discontinue to use the recommender system before the latter reaches acceptable performance. Trust networks alleviate cold-start issues by virtue of comparatively high network connectivity. Neighborhood formation hence becomes practicable even for users that explicitly trust one person only, taking into account an abundant transitive trust closure (see Section 9.5.1.1 for details).

Note that when computing neighborhoods based upon types of social relationships other than trust, e.g., geographical proximity, acquaintanceship, etc., the benefits given above may not become fully exploited.

9.4   Related Work

Decentralized recommenders have been proposed in the past already. Foner [1997, 1999] has been the first to conceive of decentralized recommender systems, pursuing an agent-based approach to the matchmaking of like-minded peers. Similarity is determined by means of feature extraction in documents, e.g., electronic mails, articles, and so forth. Olsson [1998, 2003] builds upon Foner's work and proposes another multi-agent system that addresses the issue of peers self-organizing into similarity-based clusters without central control. In both systems, the amount of messages routed through the network may prove problematic, creating a severe bottleneck. Kinateder and Rothermel [2003] propose an architecture for reputation-based matchmaking that is similar to Foner's and Olsson's approach. However, no empirical evaluations have been provided so far.

Miller [2003] explores various matchmaking techniques known from research in peer-to-peer networks, e.g., the Gnutella protocol for random discovery, Chord, and transitive traversal. Similarities between peers are computed based upon item-based and user-based CF. Hence, the sparsity issue is not part of Miller's investigations.

A taxonomization of decentralized recommender architectures along various dimensions is given by Sarwar [2001].

Systems using trust for recommendation making are still rare. Mui et al [2001] propose *collaborative sanctioning* as suitable surrogate for collaborative filtering. The idea is to make recommendations based on user *reputation* rather than user similarity. Montaner [2003] uses *trust* rather than *reputation*[1] in his multi-agent framework. Trust is regarded as direct, non-propagatable and a mere consequence of similarity. Massa and Avesani [2004] present a trust network-based recommender system, making use of simple propagation schemes. Initial results appear promising, though, at the time of this writing, their architecture still undergoes several adaptations for decentralized scenarios.

9.5 Framework Components

The following subsections and paragraphs briefly outline our decentralized, trust-based recommender system's core constituents. The information model assumed represents the union of the filtering model given in Section 3.3.1, and the trust model from Section 7.3.2.1, renaming trust functions $W_i(a_j)$ to $t_i(a_j)$ for convenience and ease of reading. Consequently, the underlying model features agents $a_i \in A$, products $b_k \in B$, implicit ratings $R_i \subseteq B$, taxonomy C and descriptors $f : B \to 2^D$, and explicit trust functions $t_i : A \to [-1, +1]^\perp$.

9.5.1 Trust-Based Neighborhood Formation

The computation of trust-based neighborhoods is one pivotal pillar of our approach. Clearly, neighborhoods are subjective, reflecting every agent a_i's very beliefs about the accorded trustworthiness of immediate peers.

9.5.1.1 Network Connectivity

However, as has been indicated before, trust functions t_i assigning *explicit* trust ratings are generally sparse. When also taking into account *indirect* trust relationships, thus exploiting the "conditional transitivity" property of trust [Abdul-Rahman and Hailes, 1997], the assembly of neighborhoods that contain M most trustworthy peers becomes possible even for larger M, e.g., $M \geq 50$. Figure 9.1 backs our hypothesis, analyzing the connectivity of 793 users from the All Consuming community (see Section 3.4.1). The chart shows the number of users, indicated on the y-axis, who satisfy the minimum neighborhood size criterion given along the x-axis. For

---

[1] See Section 7.3.1 for distinguishing both concepts from each other.

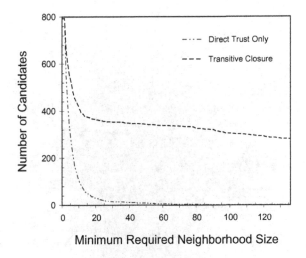

Fig. 9.1 Reach of direct trust versus transitive closure

instance, while 49 people have issued 15 or more *direct* trust statements, 374 users are able to reach 15 or more peers when also considering the *transitive closure* of trust relationships. While the trust outdegree curve decays rapidly, the transitive closure curve's fallout decelerates drastically as the number of candidate persons drops below 400, thus revealing the presence of one highly connected trust cluster (see Figure 9.2)[2].

The above result relates to the classical theorem on random graphs [Erdős and Rényi, 1959].[3] Therein, Erdős and Rényi have proved that in large graphs $G = (V, E)$, assuming E randomly assigned, the probability of getting a single gigantic component jumps from zero to one as E/V increases beyond the critical value 0.5. However, Erdős and Rényi have supposed *undirected* graphs, in contrast to our assumption of *directed* trust relationships.

Massa and Bhattacharjee [2004] have conducted experiments on top of the well-known Epinions rating community, revealing that "trust-aware techniques can produce trust scores for very high numbers of peers". Neighborhood formation thus becomes facile to achieve when considering reachability of nodes via trust paths.

---

[2] The network has been visualized with our trust testbed viewer, presented in Section 7.4.2.9.

[3] Watts and Strogatz [1998] have shown that social networks exhibit diverse "small-world" properties that make them different from random graphs, such as high clustering coefficients $C(p)$. Barabási and Albert [1999] have investigated further distinctive features, such as the scale-free nature of social networks, which is not present in random graphs. Even though, the afore-mentioned theorem holds for random graphs and social networks alike.

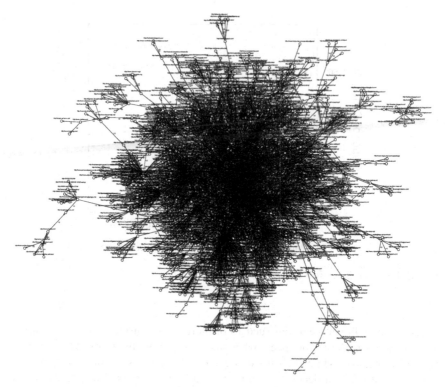

Fig. 9.2 All Consuming's largest trust cluster

9.5.1.2   Trust Propagation Models

Trust-based neighborhood computation for a_i, using those "trust-aware techniques"
mentioned by Massa, implies *deriving* trust values for peers a_j not directly trusted
by a_i, but one of the persons the latter agent trusts directly or indirectly. The trust
network's high connectivity allows assembling top-M trusted neighborhoods with
potentially large M.

Numerous scalar metrics [Beth et al, 1994; Levien and Aiken, 1998] have been
proposed for computing trust between two given individuals a_i and a_j. We hereby
denote computed trust weights by $t_i^c(a_j)$ as opposed to explicit trust $t_i(a_j)$. However,
our approach requires metrics that compute top-M *nearest trust neighbors*, and not
evaluate trust values for any two given agents. We hence opt for *local group trust
metrics* (see Chapter 7), e.g., Advogato and Appleseed. Advogato, Levien's well-
known local group trust metric, can only make *boolean* decisions with respect to
trustworthiness, simply classifying agents into trusted and non-trusted ones.

Appleseed, on the other hand, allows more fine-grained analysis, assigning
continuous trust weights for peers within trust computation range. Rankings thus
become feasible. The latter metric operates on partial trust graph information,

exploring the social network within predefined ranges only and allowing the neighborhood formation process to retain scalability. High ranks are accorded to trustworthy peers, i.e., those agents which are largely trusted by others with high trustworthiness. These ranks are used later on for selecting agents deemed suitable for making recommendations.

9.5.2 Measuring User Similarity and Product-User Relevance

Trust allows selecting peers with overall above-average interest similarity (see Chapter 8). However, for each active user a_i, some highly trusted peers a_j having completely opposed interests generally exist. The proposition that interpersonal attraction, and hence trust, implies attitudinal similarity does not always hold true. Supplementary filtering, preferably content-based, e.g., considering a_i's major fields of interest, thus becomes indispensable.

For this purpose, we apply taxonomy-driven methods to likewise compute user similarity $c(a_i, a_j)$ and product-user relevance $c_b(a_i, b_k)$, according to the computational schemes given in Section 3.3.3.1 and 3.3.4, respectively. These metrics have been designed with decentralized scenarios in mind, for common filtering metrics based upon rating vector similarity (see Section 2.3.2) tend to fail in these settings [Massa and Bhattacharjee, 2004], owing to information sparseness implied by virtually unconstrained product sets and sparse, largely implicit, rating information.

9.5.3 Recommendation Generation

Candidate recommendations of products b_k for the active user a_i are taken from the set of products that a_i's top-M neighbors have implicitly rated, discounting those products that a_i already knows. We obtain set B_i of candidate products. Next, all $b_k \in B_i$ need to be weighted according to their *relevance* for a_i. Relevance $w_i(b_k)$ hereby depends on two factors:

- **Computed trust weights** $t_i^c(a_j)$ **of peers** a_j **mentioning** b_k. Trust-based neighborhood formation supersedes finding nearest neighbors based upon interest similarity. Likewise, similarity ranks $c(a_i, a_j)$ become substituted by trust weights $t_i^c(a_j)$ for computing the predicted relevance of a_j for a_i.
- **Content-based relevance** $c_b(a_i, b_k)$ **of product** b_k **for user** a_i. Besides mere trustworthiness of peers a_j rating product b_k, the content-based relevance of b_k for the active user a_i is likewise important. For example, one may consider the situation where even close friends recommend products not fitting our interest profile at all.

We then define relevance $w_i(b_k)$ of product b_k for the active user a_i as follows, borrowing from Equation 3.6:

$$w_i(b_k) = \frac{q \cdot c_b(a_i, b_k) \cdot \sum_{a_j \in A_i(b_k)} \rho(a_i, a_j)}{|A_i(b_k)| + \Upsilon_R}, \tag{9.1}$$

where

$$A_i(b_k) = \{a_j \in \text{prox}(a_i) \,|\, b_k \in R_j\}$$

and

$$q = (1.0 + |f(b_k)| \cdot \Gamma_T)$$

In accordance with Section 3.3.4, $\text{prox}(a_i)$ denotes a_i's neighborhood, Γ_T and Υ_R represent those fine-tuning parameters introduced therein. Function $\rho(a_i, a_j)$ gives a_j's *significance* for a_i. In Equation 3.6, the latter parameter has been instantiated with the taxonomy-based user-user similarity weight $c(a_i, a_j)$.

Since we now suppose trust-based neighborhoods, $\rho(a_i, a_j) := t_i^c(a_j)$ holds.

9.6   Offline Experiments and Evaluation

The following sections present empirical results obtained from evaluating our trust-based approach for decentralized social filtering. Again, we gathered information from the All Consuming online community featuring both trust network information and product rating data. Our analysis mainly focused on pinpointing the impact that latent information kept within the trust network, namely positive interactions between interpersonal trust and attitudinal similarity (see Chapter Interpersonal-Trust-and-Similarity), may have on recommendation quality. We performed empirical offline evaluations, applying metrics introduced and used before, e.g., precision/recall according to Sarwar's definition [Sarwar et al, 2000b], and Breese score (see Section 2.4.1.2).

9.6.1   Dataset Acquisition

Currently, few online communities provide both trust *and* product rating information. To our best knowledge, Epinions and All Consuming count among the only prospective candidates. Unfortunately, Epinions has two major drawbacks that are highly pernicious for our purposes. First, owing to an immense product range diversity, most ratable products lack content meta-information. Taxonomy-based filtering thus becomes unfeasible. Second, rating information sparseness is beyond measure. For instance, Massa and Bhattacharjee [2004] have pointed out that only 8.34% of all ratable products have 10 or more reviews.

We therefore opted for the All Consuming community, which has its product range thoroughly confined to the domain of books. For the experiments at hand, we performed a third crawl, following those described in Chapter 8 and refTaxonomy-driven-Filtering. Launched on May 10, 2004, the community crawl garnered information about 3,441 users, mentioning 10,031 distinct book titles in 15,862 implicit book ratings. The accompanying trust network consisted of 4,282 links. For 9,374 of all 10,031 books, 31,157 descriptors pointing to Amazon.com's book taxonomy were found. Book ratings referring to one of those 6,55% of books not having valid taxonomic content descriptors were discarded.

One can see that using the All Consuming dataset only partially exploits functionalities our trust-based recommender system is able to unfold. For instance, the
Appleseed trust metric has been conceived with *continuous* trust and distrust statements in mind, whereas All Consuming only offers statements of *full trust*.

9.6.2 Evaluation Framework

The principal objective of our evaluation was to match the trust-based neighborhood formation scheme against other, more common approaches. Hereby, all benchmark systems were devised according to the same algorithmic clockwork, i.e., based
upon the recommendation generation framework defined in Equation 9.1. Their only
difference refers to the kind of neighborhood formation, depending on function
$\rho(a_i, a_j)$, which identifies the *relevance* of peers a_j for the active user a_i. Besides
the trust-based recommender described in Section 9.5.3, the following two recommender setups have been used for experimentation:

- **Advanced hybrid approach.** Hybrid filtering likewise exploits content-based
  and collaborative filtering facilities. Designed to eliminate intrinsic drawbacks
  of both mentioned types, this approach currently represents the most promising
  paradigm for crafting state-of-the-art recommender systems. The hybrid recommender we propose features *similarity-based* neighborhood formation, requiring $\rho(a_i, a_j) := c(a_i, a_j)$. Consequently, aside from diversification factor Θ_F, the
  approach is identical to the taxonomy-driven filtering framework presented in
  Chapter 3. Therein, we have also substantiated its superior performance over
  common benchmark recommender systems (see Section 3.4.2.4).

  However, note that this recommender's applicability is largely restricted to
  centralized scenarios only, necessitating similarity computations $c(a_i, a_j)$ for *all
  pairs* $(a_i, a_j) \in A \times A$.

- **Purely content-based filter.** Purely content-driven recommender systems *ignore* aspects of collaboration among peers and focus on content-based information only. We simulate one such recommender by supposing $\rho(a_i, a_j) :=
  \text{rnd}_{[0,1]}(a_i, a_j)$, where function $\text{rnd}_{[0,1]} : A \times A \rightarrow [0, 1]$ randomly assigns rele-
  vance weights to pairs of agents. Neighborhood formation thus amounts to an
  arbitrary sampling of users, devoid of meaningful similarity criteria. Discard-
  ing collaboration, generated recommendations are not subject to mere random,
  though. They rather depend on product features, i.e., measure $c_b(a_i, b_k)$, only.
  Hence this recommender's purely content-based nature.

Past efforts have shown that intelligent hybrid approaches tend to outperform purely
content-based ones [Huang et al, 2002; Pazzani, 1999]. We are particularly interested in beneficial ramifications resulting from trust-based neighborhood formation
as opposed to random neighborhoods. Supposing that latent semantic information
about interpersonal trust and its positive association with attitudinal similarity, en-
dogenous to the very network, has forged sufficiently strong bonds, we conjecture
that the overall recommendation quality of our trust-based approach surpasses filtering based upon content only.

9.6.2.1 Experiment Setup

The evaluation framework we established intends to compare the *utility* of recommendation lists generated by all three recommenders and roughly complies with the framework proposed in Chapter 3. Measurement is achieved by applying precision, recall, and Breese's half-life utility metric (see Section 2.4.1.2).

For cross-validation, we selected all users a_i with more than five ratings and discarded those having fewer, owing to the fact that reasonable recommendations are beyond feasibility for these cases. Moreover, users having low trust connectivity were likewise discounted. Next, we applied 5-folding, performing 80/20 splits of every user a_i's implicit ratings R_i into five pairs of *training sets* R_i^x and *test sets* T_i^x.

9.6.2.2 Parameterization

For our first experiment, neighborhood formation size was set to $M = 20$, and we provided top-20 recommendations for each active user's training set R_i^x. Proximity between profiles, based upon R_i^x and the original ratings R_j of all other agents a_j, was computed anew for each training set R_i^x of a_i.

In order to promote the impact that collaboration may have on eventual recommendations, we adopted $\Upsilon_R = 2.25$, thus rewarding books occurring frequently in ratings R_j of the active user a_i's immediate neighborhood. For content-based filtering, this parameter exerts marginal influence only. Moreover, we assumed propagation factor $\kappa = 0.75$, and topic reward $\Gamma_T = 0.1$.

9.6.3 Experiments

We conducted three diverse experiments. The first compares the effects of neighborhood formation on recommendation quality when assuming raters with varying numbers of ratings. The second investigates neighborhood size sensitivity for all three candidate schemes, while the third measures overlap of neighborhoods.

9.6.3.1 Neighborhood Formation Impact

For the first experiment, performance was analyzed by computing unweighted precision and recall (see Figure 9.3), and Breese score with half-life $\alpha = 5$ and $\alpha = 10$ (see Figure 9.4). For each indicated chart, the *minimum numbers* of ratings that users were required to have issued in order to be considered for recommendation generation and evaluation are expressed by the horizontal axis. Since all users with fewer than five ratings were ignored from the outset, performance evaluations start with all users having at least five ratings. Clearly, larger x-coordinates imply less agents considered for measurement.

Remarkably, by looking at the *differences* between the curves, more important for our analysis than the very shapes, all four charts confirm our principal hypothesis that hybrid approaches outperform purely content-based ones. Hence, promoting products that like-minded agents have voted for increases recommendation quality

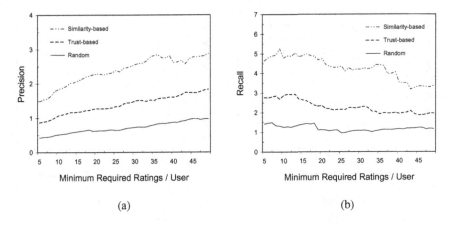

Fig. 9.3 Precision (a) and recall (b), investigating neighborhood formation

considerably. Next, we observe that our trust-based recommender significantly exceeds its purely content-based counterpart, but cannot reach the hybrid method's superior score. These results again corroborate our assumption that trust networks contain latent knowledge that reflects interest similarity between trusted agents. Clearly, trust-based neighborhood formation can only *approximate* neighborhoods assembled by means of similarity, which therefore serves as upper bound definition. However, recall that similarity-based neighborhood formation exhibits poor scalability, owing to its $O(|A|^2)$ complexity that arises from computing proximity measures $c(a_i, a_j)$ for all pairs $(a_i, a_j) \in A \times A$. Hence, this neighborhood formation scheme is not an option for decentralized recommender system infrastructures.

Trust-based clique formation, on the other hand, *does* scale and lends itself well to decentralized settings. Moreover, it bears several welcome features that similarity-based neighborhood formation does not (see Section 9.3).

The following few paragraphs investigate the *shapes* of the curves we obtained in a more fine-grained fashion. As a matter of fact, the experiment at hand corroborates our hypothesis that trust networks, in contrast to arbitrary connections between agents, bear inherent information about similarity that improves recommendation quality.

Precision

Interestingly, precision (see Figure 9.3) steadily increases even for content-based filtering. The reason for this phenomenon lies in the very nature of precision: for users a_i with test sets T_i^x smaller than the number $|P_i^x|$ of recommendations received, there is not even a chance of achieving 100% precision (see Section 3.4.2.4).

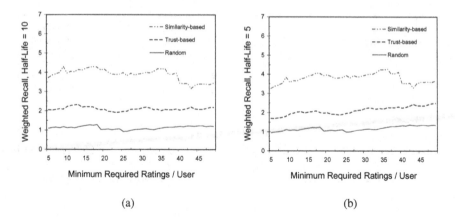

Fig. 9.4 Breese score with $\alpha = 10$ (a) and $\alpha = 5$ (b), investigating neighborhood formation

Recall

Degradation takes place for all curves when increasing x, an effect that is particularly pronounced for the hybrid recommender. Sample inspections of the All Consuming dataset suggest that infrequent raters favor bestsellers and popular books. Consequently, recommending popular books, promoted by large factor $\Upsilon_R = 2.25$, represents an appropriate guess for that particular type of users. However, when considering users possessing more refined profiles, simple "cherry picking" [Herlocker et al, 2004] does not apply anymore.

Breese Score

Scores for half-life $\alpha = 5$ and $\alpha = 10$ (see Figure 9.4) only exhibit marginal variances with respect to unweighted recall. However, degradation for increasing x becomes less pronounced when supposing lower α^4, i.e., $\alpha = 10$ and eventually $\alpha = 5$.

9.6.3.2 Neighborhood Size Sensitivity

The second experiment analyzes the impact of the neighborhood's *size* on evaluation metrics. Note that we omitted charts for weighted recall, owing to minor deviations from unweighted recall only. Figure 9.5 indicates scores for precision and recall for increasing neighborhood size $|M|$ along the horizontal axis.

Both charts exhibit similar tendencies for each neighborhood formation scheme. As it comes to similarity-based neighborhood formation, the performance of the hybrid approach steadily increases at first. Upon reaching its peak at $|M| = 25$, further increasing neighborhood size $|M|$ does not entail any gains in precision and

---

[4] Recall that unweighted recall equates Breese score with $\alpha = \infty$. .

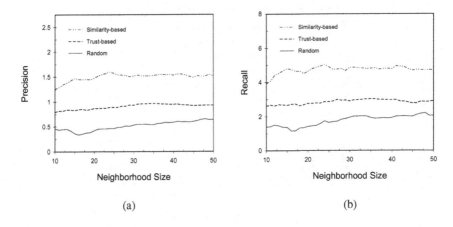

Fig. 9.5 Precision (a) and recall (b) for varying neighborhood sizes

recall anymore. This result well aligns with Sarwar's investigations for baseline collaborative filtering techniques [Sarwar et al, 2001]. Undergoing slight downward movements between $|M| = 10$ and $|M| = 15$, the content-based scheme's performance curve catches up softly. Basically, increasing the neighborhood size for the content-based filter equates to offering more candidate products[5] and easing "cherry-picking" [Herlocker et al, 2004] by virtue of large $Y_R = 2.25$.

In contrast to both other techniques, the trust-based approach shows comparatively *insensitive* to increasing neighborhood size $|M|$. As a matter of fact, its performance only marginally improves. We attribute this observation to trust's "conditional transitivity" [Abdul-Rahman and Hailes, 1997] property and Huang's investigations on transitive associations for collaborative filtering [Huang et al, 2004]: exploitation of *transitive* trust relationships, i.e., opinions of friends of friends, only works to a certain extent. However, with increasing network distance from the trust source, these peers do not satisfactorily reflect interest similarity anymore and thus represent weak predictors only. Besides empirical evidence of a positive correlation between interpersonal trust and interest similarity, as well as its positive impact on recommendation quality, we regard this aspect as one of the most important findings of our study at hand.

9.6.3.3 Neighborhood Overlap Analysis

Eventually, we compared neighborhoods formed by those three techniques. For any unordered pair $\{p,q\}$ of the three neighborhood formation techniques, we measured

---

[5] Note that only products rated by neighbors are considered for recommendation.

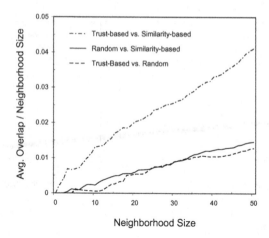

Fig. 9.6 Overlap of x-sized neighborhoods for all formation scheme pairs

the number of agents a_j occurring in *both* x-sized neighborhoods of every active user $a_i \in A$, and normalized the figure by clique size x and the number of agents $|A|$:

$$s^x(\{p,q\}) = \frac{\sum_{a_i \in A} |\operatorname{prox}_p^x(a_i) \cap \operatorname{prox}_q^x(a_i)|}{|A| \cdot x} \quad (9.2)$$

Figure 9.6 shows all three plots of $s^x(\{p,q\}), x \in [0,50]$. All curves exhibit tendencies of approximatively linear rise for increasing neighborhood size x, for the probability of overlap rises when neighborhoods become larger. Consequently, supposing clique size $x = |A|$, 100% overlap holds.

As expected, both curves displaying overlap with randomly formed neighborhoods *only marginally* differ from each other. On the other hand, the overlap between trust-based and similarity-based cliques exceeds these two baseline plots, showing that trust-based and similarity-based neighborhoods are considerably more similar to each other than pure random would allow. The above experiment again strongly corroborates our hypothesis that interpersonal trust and attitudinal similarity correlate.

9.7 Conclusion and Outlook

This chapter introduced an approach to exploit trust networks for product recommending, making use of techniques and evidence stemming from preceding chapters of this work. Superseding common collaborative approaches with trust-based filtering becomes vital when envisaging *decentralized* recommender system infrastructures, lacking central authorities.

We devised a new hybrid recommender architecture, based on the framework presented in Chapter 3, that makes use of trust-based neighborhood formation and taxonomy-driven selection of suitable products.

Moreover, we provided empirical evidence to show that network structures emanating from relationships of interpersonal trust, in contrast to random associations between users, exhibit traits of interest similarity which significantly improve recommendation quality. However, we also found that trust's tight coupling with similarity becomes lost when overly exploiting *transitive* relationships.

For our experiments, we used data from the All Consuming book reading community which offers both rating *and* trust information about its users. Note that most reputation and rating systems operating upon trust models only use *synthesized* rather than *real* trust data, therefore allowing largely limited analysis of trust semantics only.

However, we would like to base our investigations upon richer datasets in order to make our results more reliable. Unfortunately, few communities currently exist that offer accessible bulk information about both trust relationships *and* product rating data of its users. We expect this situation to change within the next years to come, owing to the steadily increasing public interest in trust networks, which is particularly promoted by the advent of weblogs and the Semantic Web. In this area, i.e., weblogs and the Semantic Web, we also see the main applicability of our proposed architecture. As the below paragraph demonstrates, current developments and trends already point into the right direction, providing an infrastructure that allows to easily leverage personal, decentralized product recommendation services into the Web.

Deployment Scenario

Referring to the information model the envisioned decentralized recommender operates upon, the model's single components can be instantiated as follows:

- **Trust networks.** FOAF (see Section 9.2.2) defines machine-readable homepages based upon RDF[6] and allows weaving acquaintance networks. Golbeck et al [2003] have proposed some modifications making FOAF support "real" trust relationships instead of mere acquaintanceship.
- **Rating information.** Moreover, FOAF networks seamlessly integrate with so-called "weblogs", which are steadily gaining momentum. These personalized online diaries are especially valuable with respect to product rating information. For instance, some crawlers extract certain hyperlinks from weblogs and analyze their makeup and content. Those links that refer to product pages from large catalogs like Amazon.com count as *implicit* votes for these goods. Mappings between hyperlinks and some sort of unique identifier are required for diverse catalogs, though. Unique identifiers exist for *some* product groups like books, which are given *International Standard Book Numbers*, i.e., ISBNs. Efforts to enhance weblogs with explicit, machine-readable rating information have

---

[6] See *http://www.w3.org/RDF/* for specifications of the RDF standard.

also been proposed and are becoming increasingly popular. For instance, BLAM! (*http://www.pmbrowser.info/hublog/*) allows creating book ratings and helps embedding these into machine-readable weblogs.

- **Classification taxonomies.** Besides user-centric information, i.e., agent a_i's trust relationships t_i and product ratings R_i, taxonomies for product classification play an important role within our approach. Luckily, these taxonomies exist for certain domains. Amazon.com defines an extensive, fine-grained and deeply-nested taxonomy for books, containing thousands of topics. More important, Amazon.com provides books with subject descriptors referring to the latter taxonomy. Similar taxonomies exist for DVDs, CDs, videos, and apparel, to name some.

  Standardization efforts for the classification of diverse kinds of consumer goods are channelled through the *United Nations Standard Products and Services Code* project (*http://www.unspsc.org/*). However, the UNSPSC's taxonomy provides much less information and nesting than, for instance, Amazon.com's taxonomy for books.

Eventually, we come to conclude that the information infrastructure required for the decentralized recommender approach described in this chapter may soon turn into reality, fostering future research on information filtering through social networks and yielding valuable large-scale evidence.

Chapter 10
Conclusion

"Too many pieces of music finish too long after the end."

– Igor Stravinsky (1882 – 1971)

10.1 Summary

Undoubtedly, recommender systems are becoming increasingly popular, owing to their versatility and their ability to reduce complexity for the human user. Current research greatly benefits from cross-fertilization, including results from other disciplines like economics (see, e.g., [Sénécal, 2003]), and behavioral sciences on the verge of HCI (see, e.g., [Swearingen and Sinha, 2001] or [Jensen et al, 2002]). The integration of interdisciplinary evidence also represents an important ingredient of this textbook, and our research on trust propagation in social networks expands the current research focus into new directions, namely that of the emerging science of social networks [Newman, 2003], and social psychology, investigating semantics of interpersonal trust.

Our research and contributions gravitate around the use of structured information that is readily available on the Web 2.0, thanks to the collective efforts of millions of humans. In this vein, we presented approaches that benefit in particular from the use of *taxonomies* and *trust* relationships.

The contributions presented in this textbook fall into mainly two categories, both being substantially different from each other:

Information Filtering. The procedure of taxonomy-driven profile creation, presented in Chapter 3, lies at the heart of our novel filtering approach and has been designed with information sparseness in mind. We integrated taxonomy-driven similarity metrics into a new framework for making product recommendations and provided comparisons with benchmark approaches.

Moreover, we distilled the topic diversification technique, an integral part of the afore-mentioned framework, and applied this particular procedure on top of conventional collaborative filtering systems in order to make top-N

C.-N. Ziegler: *Social Web Artifacts for Boosting Recommenders*, SCI 487, pp. 173–175.
DOI: 10.1007/978-3-319-00527-0_10 © Springer International Publishing Switzerland 2013

recommendation lists more meaningful (see Chapter 4). An extensive large-scale study involving more than 2, 100 human subjects and offline analyses were conducted, investigating the effects of incrementally applying topic diversification to lists generated by item-based CF and user-based CF. In addition, the study delivered a first, empirically backed argument supporting the hypothesis that "accuracy does not tell the whole story" [Cosley et al, 2002] and that there are more components to user satisfaction than pure accuracy.[1]

We then moved on in Chapter 5 to present a framework for identifying the semantic proximity of named entity pairs in an automated fashion. Herein, we also conducted empirical evaluations to benchmark our framework against human assessment. The semantic proximity computation framework serves as nucleus for Chapter 6's technology filtering method, ranking pairs of technologies according to inherent synergies.

Computational Trust. Our contributions in the field of trust and trust networks are twofold. First we introduced a new trust metric, Appleseed, which is based on spreading activation models [Quillian, 1968] known from cognitive psychology, and blends traits of PageRank [Page et al, 1998] and maximum network flow [Ford and Fulkerson, 1962]. Appleseed makes inferences in an intuitive fashion, respecting subtle semantic differences between trust and distrust, and scales to any network size. We devised Appleseed with neighborhood formation for CF in decentralized scenarios in mind.

To this end, so that trust-based neighborhoods are *meaningful* for CF applications, we conceived an evaluation framework to investigate whether interpersonal trust and interest similarity correlate, i.e., if users trusting each other were on average more similar than mere random would foretell. Again, similarity was measured by applying our taxonomy-driven similarity metric (see Chapter 3). Results obtained from an offline study on All Consuming (*http://www.allconsuming.net*) and FilmTrust (*trust.mindswap.org/FilmTrust/*) indicated that positive interactions exist (see Chapter 8), supporting the proverbial saying that "birds of a feather flock together" and levelling the ground for the application of trust-based neighborhood formation in CF systems.

Eventually, those single building bricks were put together to build a trust-based, decentralized recommender system (see Chapter 9). Note that the presented decentralized recommender's design constitutes *one possible option* among various others, giving opportunities for future research.

10.2 Discussion and Outlook

As a matter of fact, we see the underlying book's foremost strength in its versatility and variety, making contributions in diverse fields. These various contributions are not necessarily confined to the recommender system universe, but also extend to

---

[1] Some researchers, e.g., Herlocker et al [2004] and Hayes et al [2002], have raised this concern before, but have not provided any evidence whatsoever to substantiate their assertion.

other research fields. For instance, Brosowski [2004] investigates the application of Appleseed for trust-based spam filtering in electronic mails, Nejdl [2004] considers our trust metric for distributed Web search infrastructures, and Chirita et al [2004] discuss the utility of Appleseed for personalized reputation management in peer-to-peer networks.

On the other hand, the reliance of all these mosaic stones onto information rifely available on the Web 2.0, as well as their integration into a coherent framework for decentralized information filtering, gives the broader context and provides the component glue of this work.

A large portion of the research contributions outlined in this book, in particular the research on trust as well as on taxonomy-based filtering and topic diversification, has been made between 2003 and 2005. The methods of calculating semantic proximity and recommending technology synergies have been proposed between 2005 and 2009.

In the meantime, several notable events have taken place that make this books's main theme more relevant than ever before: Social networks have become even more influential and Facebook, today's by far largest network, now counts more than one billion users. That means that integrating recommender systems into those networks and exploiting their trust and friendship links is nowadays common practice. Moreover, as these social networks are looking for ways of monetization, recommender systems used for online display advertising (see, e.g., [Goldfarb and Tucker, 2011]) are omnipresent.

Yet another reason for the increased use of recommender systems in online marketing is the tendency towards *performance*-based models in advertising, as opposed to the conventional CPM[2] models: That is, publishers of display ads were paid for delivering page *impressions* rather than for the number of *clicks*, as it is the case in CPC[3] models. Now, since social networks such as Facebook have lush personal interest profiles at their disposal, expressed by each user's "likes" of brands, companies, products, and so forth, the information for targeted advertising and information filtering is all there.

After all, Web 2.0-based information filtering and recommendation generation has become reality. The book at hand thus has set some landmarks and made bold strides into the direction of more social and network-oriented recommender systems. Numerous other landmarks will follow in the near future and shape an exciting new landscape.

Welcome to this brave new world of personalization.

---

[2] Acronym for "Cost Per Mille", meaning the price to be paid for delivering 1,000 ad impressions.

[3] Abbreviation for "Cost Per Click". In this case, the advertiser has to pay if the advertisement was *clicked* rather than merely *viewed*.

References

Abdul-Rahman, A., Hailes, S.: A distributed trust model. In: New Security Paradigms Workshop, Cumbria, UK, pp. 48–60 (1997)

Abdul-Rahman, A., Hailes, S.: Supporting trust in virtual communities. In: Proceedings of the 33rd Hawaii International Conference on System Sciences, Maui, HI, USA (2000)

Aberer, K., Despotovic, Z.: Managing trust in a peer-2-peer information system. In: Paques, H., Liu, L., Grossman, D. (eds.) Proceedings of the Tenth International Conference on Information and Knowledge Management, pp. 310–317. ACM Press, Atlanta (2001)

Aggarwal, C., Wolf, J., Wu, K.L., Yu, P.: Horting hatches an egg: A new graph-theoretic approach to collaborative filtering. In: Proceedings of the Fifth ACM SIGKDD International Conference on Knowledge Discovery and Data Mining, pp. 201–212. ACM Press, San Diego (1999)

Ali, K., van Stam, W.: TiVo: Making show recommendations using a distributed collaborative filtering architecture. In: Proceedings of the 2004 ACM SIGKDD International Conference on Knowledge Discovery and Data Mining, pp. 394–401. ACM Press, Seattle (2004)

Alspector, J., Kolcz, A., Karunanithi, N.: Comparing feature-based and clique-based user models for movie selection. In: Proceedings of the Third ACM Conference on Digital Libraries, pp. 11–18. ACM Press, Pittsburgh (1998)

Armitage, P., Berry, G.: Statistical Methods in Medical Research, 3rd edn. Blackwell Science, Oxford (2001)

Avery, C., Zeckhauser, R.: Recommender systems for evaluating computer messages. Communications of the ACM 40(3), 88–89 (1997)

Baeza-Yates, R., Ribeiro-Neto, B.: Modern Information Retrieval. Addison-Wesley, Reading (1999)

Baeza-Yates, R., Ribeiro-Neto, B.: Modern Information Retrieval - the concepts and technology behind search, 2nd edn. Pearson Education Ltd., Harlow (2011)

Balabanović, M., Shoham, Y.: Fab: Content-based, collaborative recommendation. Communications of the ACM 40(3), 66–72 (1997)

Banerjee, S., Scholz, M.: Leveraging web 2.0 sources for web content classification. In: Proceedings of the 2008 IEEE/WIC/ACM International Conference on Web Intelligence, pp. 300–306 (2008)

Barabási, A.L., Albert, R.: Emergence of scaling in random networks. Science 286, 509–512 (1999)

Basilico, J., Hofmann, T.: Unifying collaborative and content-based filtering. In: Proceedings of the 21st International Conference on Machine Learning. ACM Press, Banff (2004)

Baudisch, P.: Dynamic information filtering. PhD thesis, Technische Universität Darmstadt, Darmstadt, Germany (2001)

Berscheid, E.: Interpersonal attraction. In: Gilbert, D., Fiske, S., Lindzey, G. (eds.) The Handbook of Social Psychology, 4th edn., vol. II. McGraw-Hill, New York (1998)

Beth, T., Borcherding, M., Klein, B.: Valuation of trust in open networks. In: Proceedings of the 1994 European Symposium on Research in Computer Security, Brighton, UK, pp. 3–18 (1994)

Blaze, M., Feigenbaum, J., Lacy, J.: Decentralized trust management. In: Proceedings of the 17th Symposium on Security and Privacy, pp. 164–173. IEEE Computer Society Press, Oakland (1996)

Bollegala, D., Matsuo, Y., Ishizuka, M.: Measuring semantic similarity between words using Web search engines. In: Proceedings of the 16th International Conference on World Wide Web, pp. 757–766. ACM Press (2007)

Borgida, A., Walsh, T., Hirsh, H.: Towards measuring similarity in description logics. In: Proceedings of the International Workshop on Description Logics, Ediburgh, UK, CEUR Workshop Proceedings, vol. 147 (2005)

Brafman, R., Heckerman, D., Shani, G.: Recommendation as a stochastic sequential decision problem. In: Proceedings of ICAPS 2003, Trento, Italy (2003)

Breese, J., Heckerman, D., Kadie, C.: Empirical analysis of predictive algorithms for collaborative filtering. In: Proceedings of the Fourteenth Annual Conference on Uncertainty in Artificial Intelligence, pp. 43–52. Morgan Kaufmann, Madison (1998)

Brosowski, M.: Webs of Trust in Distributed Environments: Bringing Trust to Email Communication. B.S. thesis, Information Systems Institute, University of Hannover (2004)

Budanitsky, A., Hirst, G.: Semantic distance in WordNet: An experimental, application-oriented evaluation of five measures. In: Proceedings of the Workshop on WordNet and Other Lexical Resources, Pittsburgh, PA, USA (2000)

Burgess, E., Wallin, P.: Homogamy in social characteristics. American Journal of Sociology 2(49), 109–124 (1943)

Burke, R.: Hybrid recommender systems: Survey and experiments. User Modeling and User-Adapted Interaction 12(4), 331–370 (2002)

Byrne, D.: Interpersonal attraction and attitude similarity. Journal of Abnormal and Social Psychology (62), 713–715 (1961)

Byrne, D.: The Attraction Paradigm. Academic Press, New York (1971)

Ceglowski, M., Coburn, A., Cuadrado, J.: Semantic search of unstructured data using contextual network graphs (2003)

Chen, M., Singh, J.: Computing and using reputations for internet ratings. In: Proceedings of the 3rd ACM Conference on Electronic Commerce, pp. 154–162. ACM Press, Tampa (2001)

Chen, R., Yeager, W.: Poblano: A distributed trust model for peer-to-peer networks. Tech. rep., Sun Microsystems, Santa Clara, CA, USA (2003)

Chien, S., Immorlica, N.: Semantic similarity between search engine queries using temporal correlation. In: Proceedings of the 14th International World Wide Web Conference. ACM Press, Chiba (2005)

Chirita, P.A., Nejdl, W., Schlosser, M., Scurtu, O.: Personalized reputation management in P2P networks. In: Proceedings of the ISWC 2004 Workshop on Trust, Security and Reputation, Hiroshima, Japan (2004)

Chirita, P.A., Nejdl, W., Paiu, R., Kohlschütter, C.: Using odp metadata to personlize search. In: Proceedings of the 28th International ACM SIGIR Conference on Research and Development in Information Retrieval. ACM Press, Salvador (2005)

Cimiano, P., Handschuh, S., Staab, S.: Towards the self-annotating web. In: Proceedings of the 13th International World Wide Web Conference, pp. 462–471. ACM Press, New York (2004)

Cimiano, P., Ladwig, G., Staab, S.: Gimme' the context: context-driven automatic semantic annotation with c-pankow. In: Proceedings of the 14th International World Wide Web Conference, pp. 332–341. ACM Press, Chiba (2005)

Cosley, D., Lawrence, S., Pennock, D.: REFEREE: An open framework for practical testing of recommender systems using ResearchIndex. In: 28th International Conference on Very Large Databases, pp. 35–46. Morgan Kaufmann, Hong Kong (2002)

Deshpande, M., Karypis, G.: Item-based top-n recommendation algorithms. ACM Transactions on Information Systems 22(1), 143–177 (2004)

Dumbill, E.: Finding friends with XML and RDF. IBM's XML Watch (2002)

Dwork, C., Kumar, R., Naor, M., Sivakumar, D.: Rank aggregation methods for the Web. In: Proceedings of the Tenth International Conference on World Wide Web, pp. 613–622. ACM Press, Hong Kong (2001)

Einwiller, S.: The significance of reputation and brand in creating trust between an online vendor and its customers. In: Petrovic, O., Fallenböck, M., Kittl, C. (eds.) Trust in the Network Economy, pp. 113–127. Springer, Heidelberg (2003)

Erdős, P., Rényi, A.: On random graphs. Publicationes Mathematicae 5, 290–297 (1959)

Eschenauer, L., Gligor, V., Baras, J.: On trust establishment in mobile ad-hoc networks. Tech. Rep. MS 2002-10, Institute for Systems Research, University of Maryland, MD, USA (2002)

Fagin, R., Kumar, R., Sivakumar, D.: Comparing top-k lists. In: Proceedings of the Fourteenth Annual ACM-SIAM Symposium on Discrete Algorithms, pp. 28–36. SIAM, Baltimore (2003)

Ferman, M., Errico, J., van Beek, P., Sezan, I.: Content-based filtering and personalization using structured metadata. In: Proceedings of the Second ACM/IEEE-CS Joint Conference on Digital Libraries, pp. 393–393. ACM Press, Portland (2002)

Foner, L.: Yenta: A multi-agent, referral-based matchmaking system. In: Proceedings of the First International Conference on Autonomous Agents, pp. 301–307. ACM Press, Marina del Rey (1997)

Foner, L.: Political artifacts and personal privacy: The Yenta multi-agent distributed matchmaking system. PhD thesis, Massachusetts Institute of Technology, Boston, MA, USA (1999)

Ford, L., Fulkerson, R.: Flows in Networks. Princeton University Press, Princeton (1962)

Freund, Y., Schapire, R.: Experiments with a new boosting algorithm. In: International Conference on Machine Learning, pp. 148–156. Morgan Kaufmann, Bary (1996)

Ganesan, P., Garcia-Molina, H., Widom, J.: Exploiting hierarchical domain structure to compute similarity. ACM Transactions on Information Systems 21(1), 64–93 (2003)

Gans, G., Jarke, M., Kethers, S., Lakemeyer, G.: Modeling the impact of trust and distrust in agent networks. In: Proceedings of the Third International Bi-Conference Workshop on Agent-oriented Information Systems, Montreal, Canada (2001)

Gaul, W., Schmidt-Thieme, L.: Recommender systems based on user navigational behavior in the internet. Behaviormetrika 29(1), 1–22 (2002)

Ghani, R., Fano, A.: Building recommender systems using a knowledge base of product semantics. In: Proceedings of the Workshop on Recommendation and Personalization in E-Commerce (RPEC). Springer, Malaga (2002)

Given, B.: Teaching to the Brain's Natural Learning Systems. Association for Supervision and Curriculum Development, Alexandria, VA, USA (2002)

Golbeck, J.: Computing and applying trust in Web-based social networks. PhD thesis, University of Maryland, College Park, MD, USA (2005)

Golbeck, J., Hendler, J.: Accuracy of metrics for inferring trust and reputation in Semantic Web-based social networks. In: Proceedings of the 14th International Conference on Knowledge Engineering and Knowledge Management, Northamptonshire, UK (2004)

Golbeck, J., Parsia, B., Hendler, J.: Trust networks on the Semantic Web. In: Proceedings of Cooperative Intelligent Agents, Helsinki, Finland (2003)

Goldberg, D., Nichols, D., Oki, B., Terry, D.: Using collaborative filtering to weave an information tapestry. Communications of the ACM 35(12), 61–70 (1992)

Goldberg, K., Roeder, T., Gupta, D., Perkins, C.: Eigentaste: A constant time collaborative filtering algorithm. Information Retrieval 4(2), 133–151 (2001)

Goldfarb, A., Tucker, C.: Online display advertising: Targeting and obtrusiveness. Marketing Science 30(3), 389–404 (2011)

Good, N., Schafer, B., Konstan, J., Borchers, A., Sarwar, B., Herlocker, J., Riedl, J.: Combining collaborative filtering with personal agents for better recommendations. In: Proceedings of the 16th National Conference on Artificial Intelligence and Innovative Applications of Artificial Intelligence, pp. 439–446. American Association for Artificial Intelligence, Orlando (1999)

Guha, R.: Open rating systems. Tech. rep., Stanford Knowledge Systems Laboratory, Stanford, CA, USA (2003)

Guha, R., Kumar, R., Raghavan, P., Tomkins, A.: Propagation of trust and distrust. In: Proceedings of the Thirteenth International World Wide Web Conference. ACM Press, New York (2004)

Halpin, H., Robu, V., Shepherd, H.: The complex dynamics of collaborative tagging. In: Proceedings of the 16th International World Wide Web Conference, pp. 211–220 (2007)

Hartmann, K., Strothotte, T.: A spreading activation approach to text illustration. In: Proceedings of the 2nd International Symposium on Smart Graphics, pp. 39–46. ACM Press, Hawthorne (2002)

Hayes, C., Cunningham, P.: Context-boosting collaborative recommendations. Knowledge-Based Systems 17(2-4), 131–138 (2004)

Hayes, C., Massa, P., Avesani, P., Cunningham, P.: An online evaluation framework for recommender systems. In: Proceedings of the Workshop on Personalization and Recommendation in E-Commerce (RPEC). Springer, Malaga (2002)

Heider, F.: The Psychology of Interpersonal Relations. Wiley, New York (1958)

Herlocker, J., Konstan, J., Borchers, A., Riedl, J.: An algorithmic framework for performing collaborative filtering. In: Proceedings of the 22nd Annual International ACM SIGIR Conference on Research and Development in Information Retrieval, pp. 230–237. ACM Press, Berkeley (1999)

Herlocker, J., Konstan, J., Riedl, J.: Explaining collaborative filtering recommendations. In: Proceedings of the 2000 ACM Conference on Computer-Supported Cooperative Work, Philadelphia, PA, USA, pp. 241–250 (2000)

Herlocker, J., Konstan, J., Riedl, J.: An empirical analysis of design choices in neighborhood-based collaborative filtering algorithms. Information Retrieval 5(4), 287–310 (2002)

Herlocker, J., Konstan, J., Terveen, L., Riedl, J.: Evaluating collaborative filtering recommender systems. ACM Transactions on Information Systems 22(1), 5–53 (2004)

Holland, P., Leinhardt, S.: Some evidence on the transitivity of positive interpersonal sentiment. American Journal of Sociology 77, 1205–1209 (1972)

Housely, R., Ford, W., Polk, W., Solo, D.: Internet X.509 public key infrastructure. Internet Engineering Task Force RFC 2459 (1999)

Huang, Z., Chung, W., Ong, T.H., Chen, H.: A graph-based recommender system for digital library. In: Proceedings of the Second ACM/IEEE-CS Joint Conference on Digital Libraries, pp. 65–73. ACM Press, Portland (2002)

Huang, Z., Chen, H., Zeng, D.: Applying associative retrieval techniques to alleviate the sparsity problem in collaborative filtering. ACM Transactions on Information Systems 22(1), 116–142 (2004)

Huston, T., Levinger, G.: Interpersonal attraction and relationships. Annual Review of Psychology 29, 115–156 (1978)

Jäschke, R., Marinho, L., Hotho, A., Schmidt-Thieme, L., Stumme, G.: Tag recommendations in folksonomies. In: Kok, J.N., Koronacki, J., Lopez de Mantaras, R., Matwin, S., Mladenič, D., Skowron, A. (eds.) PKDD 2007. LNCS (LNAI), vol. 4702, pp. 506–514. Springer, Heidelberg (2007)

Jensen, C., Davis, J., Farnham, S.: Finding others online: Reputation systems for social online spaces. In: Proceedings of the SIGCHI Conference on Human Factors in Computing Systems, pp. 447–454. ACM Press, Minneapolis (2002)

Jiang, J., Conrath, D.: Semantic similarity based on corpus statistics and lexical taxonomy. In: Proceedings of the International Conference on Research in Computational Linguistics, Taiwan (1997)

Jones, E., Bell, L., Aronson, E.: The reciprocation of attraction from similar and dissimilar others. In: McClintock, C. (ed.) Experimental Social Psychology. Holt, Rinehart, and Winston, New York (1972)

Jøsang, A., Gray, E., Kinateder, M.: Analysing topologies of transitive trust. In: Proceedings of the Workshop of Formal Aspects of Security and Trust, Pisa, Italy (2003)

Kamvar, S., Schlosser, M., Garcia-Molina, H.: The Eigentrust algorithm for reputation management in P2P networks. In: Proceedings of the 12th International Conference on World Wide Web, pp. 640–651. ACM Press, Budapest (2003)

Karypis, G.: Evaluation of item-based top-n recommendation algorithms. In: Proceedings of the Tenth ACM CIKM International Conference on Information and Knowledge Management, pp. 247–254. ACM Press, Atlanta (2001)

Kautz, H., Selman, B., Shah, M.: Referral Web: Combining social networks and collaborative filtering. Communications of the ACM 40(3), 63–65 (1997)

Kinateder, M., Pearson, S.: A privacy-enhanced peer-to-peer reputation system. In: Bauknecht, K., Min Tjoa, A., Quirchmayr, G. (eds.) EC-Web 2003. LNCS, vol. 2738, pp. 206–216. Springer, Heidelberg (2003)

Kinateder, M., Rothermel, K.: Architecture and algorithms for a distributed reputation system. In: Nixon, P., Terzis, S. (eds.) iTrust 2003. LNCS, vol. 2692, pp. 1–16. Springer, Heidelberg (2003)

Kleinberg, J.: Authoritative sources in a hyperlinked environment. Journal of the ACM 46(5), 604–632 (1999)

Konstan, J.: Introduction to recommender systems: Algorithms and evaluation. ACM Transactions on Information Systems 22(1), 1–4 (2004)

Konstan, J., Miller, B., Maltz, D., Herlocker, J., Gordon, L., Riedl, J.: GroupLens: Applying collaborative filtering to usenet news. Communications of the ACM 40(3), 77–87 (1997)

Kummamuru, K., Lotlikar, R., Roy, S., Singal, K., Krishnapuram, R.: A hierarchical monothetic document clustering algorithm for summarization and browsing search results. In: Proceedings of the 13th International Conference on World Wide Web, pp. 658–665. ACM Press, New York (2004)

Lam, S., Riedl, J.: Shilling recommender systems for fun and profit. In: Proceedings of the 13th International Conference on World Wide Web, pp. 393–402. ACM Press, New York (2004)

Lam, W., Mukhopadhyay, S., Mostafa, J., Palakal, M.: Detection of shifts in user interests for personalized information filtering. In: Proceedings of the 19th ACM SIGIR Conference on Research and Development in Information Retrieval, pp. 317–325. ACM Press, Zürich (1996)

Lang, K.: NewsWeeder: Learning to filter netnews. In: Proceedings of the 12th International Conference on Machine Learning, pp. 331–339. Morgan Kaufmann, San Mateo (1995)

Levien, R.: Attack-resistant trust metrics. PhD thesis, University of California at Berkeley, Berkeley, CA, USA (2004) (to appear)

Levien, R., Aiken, A.: Attack-resistant trust metrics for public key certification. In: Proceedings of the 7th USENIX Security Symposium, San Antonio, TX, USA (1998)

Levien, R., Aiken, A.: An attack-resistant, scalable name service. Draft submission to the Fourth International Conference on Financial Cryptography (2000)

Lewicki, R., McAllister, D., Bies, R.: Trust and distrust: New relationships and realities. Academy of Management Review 23(12), 438–458 (1998)

Li, Y., Bandar, Z., McLean, D.: An approach for measuring semantic similarity between words using multiple information sources. IEEE Transactions on Knowledge and Data Engineering 15(4), 871–882 (2003)

Lin, D.: An information-theoretic definition of similarity. In: Proceedings of the Fifteenth International Conference on Machine Learning, pp. 296–304. Morgan Kaufmann, San Francisco (1998)

Linden, G., Smith, B., York, J.: Amazon.com recommendations: Item-to-item collaborative filtering. IEEE Internet Computing 4(1) (2003)

Luhmann, N.: Trust and Power. Wiley, Chichester (1979)

Lutz, C., Wolter, F., Zakharyaschev, M.: Resasoning about concepts and similarity. In: Calvanese, D., Giacomo, G.D., Franconi, E. (eds.) Description Logics, Rome, Italy. CEUR Workshop Proceedings, vol. 81 (2003)

Maguitman, A., Menczer, F., Roinestad, H., Vespignani, A.: Algorithmic detection of semantic similarity. In: Proceedings of the 14th International World Wide Web Conference, pp. 107–116. ACM Press, Chiba (2005)

Malone, T., Grant, K., Turbak, F., Brobst, S., Cohen, M.: Intelligent information-sharing systems. Communications of the ACM 30(5), 390–402 (1987)

Manning, C., Schütze, H.: Foundations of statistical natural language processing. MIT Press (2001)

Marsh, S.: Formalising trust as a computational concept. PhD thesis, Department of Mathematics and Computer Science, University of Stirling, Stirling, UK (1994a)

Marsh, S.: Optimism and pessimism in trust. In: Ramirez, J. (ed.) Proceedings of the Ibero-American Conference on Artificial Intelligence. McGraw-Hill, Caracas (1994b)

Massa, P., Avesani, P.: Trust-aware collaborative filtering for recommender systems. In: Meersman, R., Tari, Z. (eds.) OTM 2004. LNCS, vol. 3290, pp. 492–508. Springer, Heidelberg (2004)

Massa, P., Bhattacharjee, B.: Using trust in recommender systems: An experimental analysis. In: Jensen, C., Poslad, S., Dimitrakos, T. (eds.) iTrust 2004. LNCS, vol. 2995, pp. 221–235. Springer, Heidelberg (2004)

Maurer, U.: Modelling a public key infrastructure. In: Bertino, E., Kurth, H., Martella, G., Montolivo, E. (eds.) ESORICS 1996. LNCS, vol. 1146, pp. 325–350. Springer, Heidelberg (1996)

McKnight, H., Chervany, N.: The meaning of trust. Tech. Rep. MISRC 96-04, Management Information Systems Research Center, University of Minnesota, MN, USA (1996)

McNee, S., Albert, I., Cosley, D., Gopalkrishnan, P., Lam, S., Rashid, A., Konstan, J., Riedl, J.: On the recommending of citations for research papers. In: Proceedings of the 2002 ACM Conference on Computer-Supported Cooperative Work, pp. 116–125. ACM Press, New Orleans (2002)

Melville, P., Mooney, R., Nagarajan, R.: Content-boosted collaborative filtering for improved recommendations. In: Eighteenth National Conference on Artificial Intelligence, pp. 187–192. American Association for Artificial Intelligence, Edmonton (2002)

Middleton, S., De Roure, D., Shadbolt, N.: Capturing knowledge of user preferences: Ontologies in recommender systems. In: Proceedings of the First International Conference on Knowledge Capture, Victoria, Canada (2001)

Middleton, S., Alani, H., Shadbolt, N., De Roure, D.: Exploiting synergy between ontologies and recommender systems. In: Proceedings of the WWW2002 International Workshop on the Semantic Web, Maui, HI, USA. CEUR Workshop Proceedings, vol. 55 (2002)

Middleton, S., Shadbolt, N., De Roure, D.: Ontological user profiling in recommender systems. ACM Transactions on Information Systems 22(1), 54–88 (2004)

Milgram, S.: The small world problem. In: Sabini, J., Silver, M. (eds.) The Individual in a Social World - Essays and Experiments, 2nd edn. McGraw-Hill, New York (1992)

Miller, B.: Toward a personalized recommender system. PhD thesis, University of Minnesota, Minneapolis, MA, USA (2003)

Miller, G.: Wordnet: A lexical database for english. Communications of the ACM 38(11), 39–41 (1995)

Miller, G., Charles, W.: Contextual correlates of semantic similarity. Language and Cognitive Processes 6(1), 1–28 (1991)

Miyahara, K., Pazzani, M.: Collaborative filtering with the simple bayesian classifier. In: Proceedings of the 6th Pacific Rim International Conference on Artificial Intelligence, Melbourne, Australia, pp. 679–689 (2000)

Montaner, M.: Collaborative recommender agents based on case-based reasoning and trust. PhD thesis, Universitat de Girona, Girona, Spain (2003)

Montaner, M., López, B., de la Rosa, J.L.: Opinion-based filtering through trust. In: Klusch, M., Ossowski, S., Shehory, O. (eds.) CIA 2002. LNCS (LNAI), vol. 2446, pp. 164–178. Springer, Heidelberg (2002)

Mui, L., Szolovits, P., Ang, C.: Collaborative sanctioning: Applications in restaurant recommendations based on reputation. In: Proceedings of the Fifth International Conference on Autonomous Agents, pp. 118–119. ACM Press, Montreal (2001)

Mui, L., Mohtashemi, M., Halberstadt, A.: A computational model of trust and reputation. In: Proceedings of the 35th Hawaii International Conference on System Sciences, Big Island, HI, USA, pp. 188–196 (2002)

Mukherjee, R., Dutta, P., Sen, S.: MOVIES2GO: A new approach to online movie recommendation. In: Proceedings of the IJCAI Workshop on Intelligent Techniques for Web Personalization, Seattle, WA, USA (2001)

Nejdl, W.: How to build Google2Google: An incomplete receipe (Invited talk). In: McIlraith, S.A., Plexousakis, D., van Harmelen, F. (eds.) ISWC 2004. LNCS, vol. 3298, pp. 1–5. Springer, Heidelberg (2004)

Newcomb, T.: The Acquaintance Process. Holt, Rinehart, and Winston, New York (1961)

Newman, M.: The structure and function of complex networks. SIAM Review 45(2), 167–256 (2003)

Nichols, D.: Implicit rating and filtering. In: Proceedings of the Fifth DELOS Workshop on Filtering and Collaborative Filtering, pp. 31–36. ERCIM, Budapest (1998)

Obrst, L., Liu, H., Wray, R.: Ontologies for corporate Web applications. AI Magazine 24(3), 49–62 (2003)

Olsson, T.: Decentralized social filtering based on trust. Working Notes of the AAAI 1998 Recommender Systems Workshop, Madison, WI, USA (1998)

Olsson, T.: Bootstrapping and decentralizing recommender systems. PhD thesis, Uppsala University, Uppsala, Sweden (2003)

O'Mahony, M., Hurley, N., Kushmerick, N., Silvestre, G.: Collaborative recommendation: A robustness analysis. ACM Transactions on Internet Technology 4(3) (2004)

Oztekin, U., Karypis, G., Kumar, V.: Expert agreement and content-based reranking in a meta search environment using Mearf. In: Proceedings of the Eleventh International Conference on World Wide Web, pp. 333–344. ACM Press, Honolulu (2002)

Page, L., Brin, S., Motwani, R., Winograd, T.: The PageRank citation ranking: Bringing order to the Web. Tech. rep., Stanford Digital Library Technologies Project (1998)

Pazzani, M.: A framework for collaborative, content-based and demographic filtering. Artificial Intelligence Review 13(5-6), 393–408 (1999)

Pescovitz, D.: The best new technologies of 2003. Business 20(11) (2003)

Pretschner, A., Gauch, S.: Ontology-based personalized search. In: Proceedings of the 11th IEEE International Conference on Tools with Artificial Intelligence, Chicago, IL, USA, pp. 391–398 (1999)

Quillian, R.: Semantic memory. In: Minsky, M. (ed.) Semantic Information Processing, pp. 227–270. MIT Press, Boston (1968)

Rapoport, A.: Mathematical models of social interaction. In: Luce, D., Bush, R., Galanter, E. (eds.) Handbook of Mathematical Psychology, vol. 2. Wiley, New York (1963)

Reiter, M., Stubblebine, S.: Path independence for authentication in large-scale systems. In: Proceedings of the ACM Conference on Computer and Communications Security, pp. 57–66. ACM Press, Zürich (1997a)

Reiter, M., Stubblebine, S.: Toward acceptable metrics of authentication. In: Proceedings of the IEEE Symposium on Security and Privacy, pp. 10–20. IEEE Computer Society Press, Oakland (1997b)

Resnick, P., Varian, H.: Recommender systems. Communications of the ACM 40(3), 56–58 (1997)

Resnick, P., Iacovou, N., Suchak, M., Bergstorm, P., Riedl, J.: GroupLens: An open architecture for collaborative filtering of netnews. In: Proceedings of the ACM 1994 Conference on Computer-Supported Cooperative Work, pp. 175–186. ACM, Chapel Hill (1994)

Resnik, P.: Using information content to evaluate semantic similarity in a taxonomy. In: Proceedings of the Fourteenth International Joint Conference on Artificial Intelligence, Montreal, Canada, pp. 448–453 (1995)

Resnik, P.: Semantic similarity in a taxonomy: An information-based measure and its application to problems of ambiguity in natural language. Journal of Artificial Intelligence Research 11, 95–130 (1999)

Richardson, M., Agrawal, R., Domingos, P.: Trust management for the Semantic Web. In: Proceedings of the Second International Semantic Web Conference, Sanibel Island, FL, USA (2003)

van Rijsbergen, K.: Information Retrieval. Butterworths, London (1975)

Ritzer, G., Jurgenson, N.: Production, consumption, prosumption: The nature of capitalism in the age of the digital prosumer. Journal of Consumer Culture 10(1), 13–36 (2010)

Rubenstein, H., Goodenough, J.: Contextual correlates of synonymy. Communications of the ACM 8(10), 627–633 (1965)

Sahami, M., Heilman, T.: A Web-based kernel function for measuring the similarity of short text snippets. In: Proceedings of the 15th International Conference on World Wide Web, pp. 377–386. ACM Press, Edinburgh (2006)

Sankaralingam, K., Sethumadhavan, S., Browne, J.: Distributed PageRank for P2P systems. In: Proceedings of the Twelfth International Symposium on High Performance Distributed Computing, Seattle, WA, USA (2003)

Sarwar, B.: Sparsity, scalability, and distribution in recommender systems. PhD thesis, University of Minnesota, Minneapolis, MA, USA (2001)

Sarwar, B., Konstan, J., Borchers, A., Herlocker, J., Miller, B., Riedl, J.: Using filtering agents to improve prediction quality in the GroupLens research collaborative filtering system. In: Proceedings of the 1998 ACM Conference on Computer-Supported Cooperative Work, pp. 345–354. ACM Press, Seattle (1998)

Sarwar, B., Karypis, G., Konstan, J., Riedl, J.: Analysis of recommendation algorithms for e-commerce. In: Proceedings of the 2nd ACM Conference on Electronic Commerce, pp. 158–167. ACM Press, Minneapolis (2000a)

Sarwar, B., Karypis, G., Konstan, J., Riedl, J.: Application of dimensionality reduction in recommender systems. In: ACM WebKDD Workshop, Boston, MA, USA (2000b)

Sarwar, B., Karypis, G., Konstan, J., Riedl, J.: Item-based collaborative filtering recommendation algorithms. In: Proceedings of the Tenth International World Wide Web Conference, Hong Kong, China (2001)

Schafer, B., Konstan, J., Riedl, J.: Recommender systems in e-commerce. In: Proceedings of the First ACM Conference on Electronic Commerce, pp. 158–166. ACM Press, Denver (1999)

Schein, A., Popescul, A., Ungar, L., Pennock, D.: Methods and metrics for cold-start recommendations. In: Proceedings of the 25th Annual International ACM SIGIR Conference on Research and Development in Information Retrieval, pp. 253–260. ACM Press, Tampere (2002)

Sénécal, S.: Essays on the influence of online relevant others on consumers' online product choices. PhD thesis, École des Hautes Études Commerciales, Université de Montréal, Montreal, Canada (2003)

Shardanand, U., Maes, P.: Social information filtering: Algorithms for automating "word of mouth". In: Proceedings of the ACM CHI Conference on Human Factors in Computing Systems, pp. 210–217. ACM Press, Denver (1995)

Siegel, S., Castellan, J.: Non-parametric Statistics for the Behavioral Sciences, 2nd edn. McGraw-Hill, New York (1988)

Sinha, R., Swearingen, K.: Comparing recommendations made by online systems and friends. In: Proceedings of the DELOS-NSF Workshop on Personalization and Recommender Systems in Digital Libraries, Dublin, Ireland (2001)

Smith, E., Nolen-Hoeksema, S., Fredrickson, B., Loftus, G.: Atkinson and Hilgards's Introduction to Psychology. Thomson Learning, Boston (2003)

Snyder, C., Fromkin, H.: Uniqueness: The Human Pursuit of Difference. Plenum, New York (1980)

Sollenborn, M., Funk, P.: Category-based filtering and user stereotype cases to reduce the latency problem in recommender systems. In: Craw, S., Preece, A.D. (eds.) ECCBR 2002. LNCS (LNAI), vol. 2416, pp. 395–405. Springer, Heidelberg (2002)

Srikumar, K., Bhasker, B.: Personalized recommendations in e-commerce. In: Proceedings of the 5th World Congress on Management of Electronic Business, Hamilton, Canada (2004)

Swearingen, K., Sinha, R.: Beyond algorithms: An HCI perspective on recommender systems. In: Proceedings of the ACM SIGIR 2001 Workshop on Recommender Systems, New Orleans, LA, USA (2001)

Terveen, L., Hill, W.: Beyond recommender systems: Helping people help each other. In: Carroll, J. (ed.) Human-Computer Interaction in the New Millenium. Addison-Wesley, Reading (2001)

Tombs, M.: Osmotic Pressure of Biological Macromolecules. Oxford University Press, New York (1997)

Torres, R., McNee, S., Abel, M., Konstan, J., Riedl, J.: Enhancing digital libraries with Tech-Lens+. In: Proceedings of the 2004 Joint ACM/IEEE Conference on Digital Libraries, pp. 228–236. ACM Press, Tuscon (2004)

Turney, P.D.: Mining the web for synonyms: PMI-IR versus LSA on TOEFL. In: Flach, P.A., De Raedt, L. (eds.) ECML 2001. LNCS (LNAI), vol. 2167, pp. 491–502. Springer, Heidelberg (2001)

Twigg, A., Dimmock, N.: Attack-resistance of computational trust models. In: Proceedings of the Twelfth IEEE International Workshop on Enabling Technologies, Linz, Austria, pp. 275–280 (2003)

Vlachos, M., Meek, C., Vagena, Z., Gunopulos, D.: Identifying similarities, periodicities and bursts for online search queries. In: Proceedings of the 2004 ACM SIGMOD International Conference on Management of Data, pp. 131–142. ACM Press, Paris (2004)

Vogt, C., Cottrell, G.: Fusion via a linear combination of scores. Information Retrieval 1(3), 151–173 (1999)

Watts, D., Strogatz, S.: Collective dynamics of "small-world" networks. Nature 393, 440–442 (1998)

Wen, J.R., Nie, J.Y., Zhang, H.J.: Clustering user queries of a search engine. In: Proceedings of the 10th International World Wide Web Conference, pp. 162–168. ACM Press, Hong Kong (2001)

Wiese, R., Eiglsperger, M., Kaufmann, M.: yFiles: Visualization and automatic layout of graphs. In: Mutzel, P., Jünger, M., Leipert, S. (eds.) GD 2001. LNCS, vol. 2265, pp. 453–454. Springer, Heidelberg (2002)

Ziegler, C.N.: Semantic Web recommender systems. In: Lindner, W., Perego, A. (eds.) Proceedings of the Joint ICDE/EDBT Ph.D. Workshop 2004. Crete University Press, Heraklion (2004a)

Ziegler, C.N.: Semantic web recommender systems. In: Lindner, W., Fischer, F., Türker, C., Tzitzikas, Y., Vakali, A.I. (eds.) EDBT 2004. LNCS, vol. 3268, pp. 78–89. Springer, Heidelberg (2004)

Ziegler, C.N.: On propagating interpersonal trust in social networks. In: Golbeck, J. (ed.) Computing with Social Trust, 1st edn. Human-Computer Interaction Series. Springer, Heidelberg (2009)

Ziegler, C.N., Golbeck, J.: Investigating interactions of trust and interest similarity. Decision Support Systems 43(2), 460–475 (2007)

Ziegler, C.N., Jung, S.: Leveraging sources of collective wisdom on the web for discovering technology synergies. In: Proceedings of the 2006 ACM SIGIR Conference on Research and Development in Information Retrieval, pp. 548–555. ACM Press, Boston (2009)

Ziegler, C.N., Lausen, G.: Analyzing correlation between trust and user similarity in on-line communities. In: Jensen, C., Poslad, S., Dimitrakos, T. (eds.) iTrust 2004. LNCS, vol. 2995, pp. 251–265. Springer, Heidelberg (2004)

Ziegler, C.N., Lausen, G.: Paradigms for decentralized social filtering exploiting trust network structure. In: Meersman, R. (ed.) OTM 2004. LNCS, vol. 3291, pp. 840–858. Springer, Heidelberg (2004)

Ziegler, C.N., Lausen, G.: Spreading activation models for trust propagation. In: Proceedings of the IEEE International Conference on e-Technology, e-Commerce, and e-Service. IEEE Computer Society Press, Taipei (2004)

Ziegler, C.N., Lausen, G.: Propagation models for trust and distrust in social networks. Information Systems Frontiers 7(4-5), 337–358 (2005)

Ziegler, C.N., Lausen, G.: Making product recommendations more diverse. IEEE Data Engineering Bulletin 32(4), 23–32 (2009)

Ziegler, C.N., Viermetz, M.: Discovery of technology synergies through collective wisdom. In: Proceedings of the 2009 IEEE/WIC/ACM International Conference on Web Intelligence, Milan, Italy, pp. 701–706 (2009)

Ziegler, C.N., Lausen, G., Schmidt-Thieme, L.: Taxonomy-driven computation of product recommendations. In: Proceedings of the 2004 ACM CIKM Conference on Information and Knowledge Management, pp. 406–415. ACM Press, Washington, D.C (2004a)

Ziegler, C.N., Schmidt-Thieme, L., Lausen, G.: Exploiting semantic product descriptions for recommender systems. In: Proceedings of the 2nd ACM SIGIR Semantic Web and Information Retrieval Workshop, Sheffield, UK (2004b)

Ziegler, C.N., McNee, S., Konstan, J., Lausen, G.: Improving recommendation lists through topic diversification. In: Proceedings of the 14th International World Wide Web Conference. ACM Press, Chiba (2005)

Ziegler, C.N., Simon, K., Lausen, G.: Automatic computation of semantic proximity using taxonomic knowledge. In: Proceedings of the 2006 ACM CIKM Conference on Information and Knowledge Management. ACM Press, Washington, D.C (2006)

Zimmermann, P.: The Official PGP User's Guide. MIT Press, Boston (1995)

Printed in the United States
By Bookmasters